Lecture Notes in Business Information Processing 280

More information about this series at http://www.springer.com/series/7911

Patrick Marcel · Esteban Zimányi (Eds.)

Business Intelligence

6th European Summer School, eBISS 2016
Tours, France, July 3–8, 2016
Tutorial Lectures

Springer

Editors
Patrick Marcel
University of Tours
Tours
France

Esteban Zimányi
Department of Computer and Decision Engineering (CoDE)
Université Libre de Bruxelles
Brussels
Belgium

ISSN 1865-1348 ISSN 1865-1356 (electronic)
Lecture Notes in Business Information Processing
ISBN 978-3-319-61163-1 ISBN 978-3-319-61164-8 (eBook)
DOI 10.1007/978-3-319-61164-8

Library of Congress Control Number: 2017944217

Printed on acid-free paper

This Springer imprint is published by Springer Nature
The registered company is Springer International Publishing AG
The registered company address is: Gewerbestrasse 11, 6330 Cham, Switzerland

Preface

The 6th European Business Intelligence and Big Data Summer School (eBISS 2016) took place in Tours, France, in July 2016. Tutorials were given by renowned experts and covered recent and various aspects of business intelligence and big data. This volume contains the lecture notes of the summer school.

The first chapter aims at equipping the reader with a fundamental understanding of how to perform analytics on graph data. It starts by reviewing the concepts that form the common basis of declarative graph querying languages, thereby helping the reader to work with graph database systems in general. A particular focus is placed on the fundamental aspects of subgraph matching and graph transformations. These concepts are illustrated with examples in query languages such as Cypher and SPARQL. Finally, it discusses the extension of graph query languages toward declarative multidimensional queries.

The second chapter targets the readers interested in machine translation. It introduces some of the findings of translation studies to better and more accurate machine translation systems. It starts by surveying some theoretical hypotheses of translation studies, and then describes works that use standard text classification techniques to distinguish between translations and originals. Finally, it shows how language models compiled from translated texts and translation models compiled from texts translated from the source to the target can improve machine translation.

The third chapter presents an overview of pattern mining, which has been one of the most active fields in knowledge discovery in databases for the last two decades. It first introduces the concepts of language and interestingness, which rule the pattern mining process. It then reviews the main categories of extraction methods, that either enumerate all patterns whose interestingness exceeds a user-specified threshold or enumerate all the patterns whose interest is maximum. Finally, it introduces interactive pattern mining as a way to discover what is the user's interest while mining relevant patterns.

The fourth chapter discusses models, concepts, and approaches for reaching scalability and real time in big data processing and big data warehouses. It reviews the concepts of NoSQL, parallel data management systems (PDBMS), MapReduce, Spark, real time data processing, and lambda architecture in the context of scalability. It then presents three specific approaches for real time and scalability and exposes some of the major current solutions for real time scalable big data analytics.

Finally, the fifth chapter introduces the challenges around reducing the energy consumed by DBMS and more specifically data warehouses. It first presents a generic framework integrating the energy in query optimizers of DBMS hosting already-designed DW. It then discusses how energy consumption may be integrated in the logical phase of DW life cycle, and evaluates its impact on the physical phase. It concludes by presenting experiments to evaluate the effectiveness and efficiency of this framework, using PostgreSQL and Oracle DBMS.

In addition to the lectures corresponding to the chapters described, eBISS 2016 had two other lectures directly related to industry:

- Alexis Naibo from SAP, France: "The BIG Trends in BICC (BI Competency Centers)"
- Pierre Maussion and Marie Pérennès from Teradata, France: "Discovery of the Teradata Database"

These lectures have no associated chapter in this volume.

Like the fifth edition, eBISS joined forces with the Erasmus Mundus IT4BI-DC consortium and hosted its doctoral colloquium aiming at community building and promoting a corporate spirit among PhD candidates, advisors, and researchers of different organizations. The corresponding two sessions, each organized in two parallel tracks, included 17 presentations, as follows:

- Besim Bellali: ETL Design for Advanced Analytics
- Rudra Nath: Data Integration and ETL for Semantic Web
- Faisal Orakzai: Mobility Data Management and Analysis
- Ahmad Ahmedov: Analyze and Explore the Web of Data
- Muhammad Aamir Saleem: Advanced Geo-Social Analytics
- Kai Herrmann: Database Versioning
- Fawad Ali: Parallelization of User-Defined ETL Tasks in an ETL Workflow
- Ayman Al-Serafi: Discovering Semantic Metadata in the Data Lake
- Davide Frazetto: Advancing Data Analytics: A Platform for Integrated Data Management and Prescriptive Analytics
- Muhammad Idris: Active Business Intelligence Through Incremental View Maintenance on Factorized Representations
- Bhuvan Gummidi: Integrated Framework for Recruiting Moving Spatial Crowdsourcing Workers with Quality Assurance
- Rohit Kumar: Converging and Adaptive Stream Processing
- Rana Faisal Munir: Physical Design by Requirements Integration
- Elvis Koci: From Partially Structured Documents to Relations
- Gaston Bakkalian: Data Structures and Algorithms for Sequential Data Warehouse
- Sergi Nadal: Self-Optimizing Data Stream Processing
- Lawan Subba: bitmap Indexing for Big Data

We would like to thank the attendees of the summer school for their active participation, as well as the speakers and their co-authors for the high quality of their contribution in a constantly evolving and highly competitive domain. Finally, we would like to thank the reviewers for their careful evaluation of the chapters.

March 2017

Patrick Marcel
Esteban Zimányi

Organization

The 6th European Business Intelligence and Big Data Summer School (eBISS 2016) was organized by the Department of Computer and Decision Engineering (CoDE) of the Université Libre de Bruxelles, Belgium, and the Laboratoire d'Informatique of the Université François Rabelais de Tours, France.

Program Committee

Alberto Abelló	Universitat Politècnica de Catalunya, BarcelonaTech, Spain
Nacéra Bennacer	Centrale-Supélec, France
Ralf-Detlef Kutsche	Technische Universität Berlin, Germany
Patrick Marcel	Université François Rabelais de Tours, France
Esteban Zimányi	Université Libre de Bruxelles, Belgique

Additional Reviewers

Andreas Behrend	University of Bonn, Germany
Katja Hose	Aalborg University, Denmark
Iustina Ilisei	University of Wolverhampton, UK
Mehdi Kaytoue	INSA Lyon, France
Carson K. Leung	University of Manitoba, Canada
Sofian Maabout	Université de Bordeaux, France
Tilmann Rabl	Technische Universität Berlin, Germany
Prajwol Sangat	Monash University, Australia
Agata Savary	Université François Rabelais de Tours, France
David Taniar	Monash University, Australia
Robert Wrembel	Poznan University of Technology, Poland
Albrecht Zimmermann	Université de Caen, France

Sponsorship and Support

Université François Rabelais de Tours
Région Centre-Val de Loire
GdR MADICS
EACEA

Contents

Declarative Multidimensional Graph Queries

Hannes Voigt(✉)

Dresden Database Systems Group, Technische Universität Dresden,
01062 Dresden, Germany
hannes.voigt@tu-dresden.de
http://wwwdb.inf.tu-dresden.de

Abstract. Graphs have become an ubiquitous type of data, increasing the desire and need to perform analytics on graph data. In this article, we review the fundamental concepts that form the common basis of most declarative graph querying languages. The article conveys a general understanding of these concepts which will help the reader to learn a specific graph query language, rotate between graph query language, and work with graph database system in general. We also provide examples in query languages such as Cypher and SPARQL that illustrate how the discussed concept manifest in specific graph query languages. Further, we take a look at how these concepts of declarative graph query languages can be extended towards declarative multidimensional queries, to facilitate the most fundamental form of analytics on graphs in graph database systems. This helps the reader to develop a fundamental understanding of aggregating graph structure for analytical purposes.

Keywords: Graph query language · Subgraph matching · Subgraph isomorphism · Subgraph homomorphism · Graph transformation · Multidimensional graph query

1 Introduction

The mathematical concept of graph is somewhat a rediscovered old friend in the database community. Predating the relational data model, the CODASYL network data model resembles essentially graph data. In the 1980s and early 1990s, with the rise of object-oriented programming and advent of object-oriented database systems, research considered graph-based data models and graph query languages. With the continued dominance of relational DBMSs, none of these efforts got any sustainable traction in industry. In the last decade, however, the graph concept has had a considerable revival with three major trends driving it.

The first driver is the semantic web movement [1]. The idea of the semantic web gave rise to the Resource Description Framework (RDF), which structures data as a labeled graph. This propelled the publication and maintenance of thousands of open RDF datasets on the internet, most famously DBpedia [2]. It also sparked research in every corner of the database community – ranging from

P. Marcel and E. Zimányi (Eds.): eBISS 2016, LNBIP 280, pp. 1–37, 2017.
DOI: 10.1007/978-3-319-61164-8_1

works investigating the fundamental properties of query languages for labeled graphs to the design of storage structures and query engines for RDF data.

The second driver is agility with respect to the management of data. New application domains (e.g. dataspaces [3], visual data modeling and exploration [4]) as well as novel development methods [5] increased the demand for data models that are less rigid and schema-oriented but more ad-hoc and data-oriented. Graph data models typically excel in this regard as new vertices and edges can be added anytime, regardless of their properties. This propelled the proliferation of the Property Graph model and corresponding DBMSs, such as Neo4j and Apache TinkerPop Blueprints implementations. By now, also major DBMS vendors such as IBM and Oracle have put their weight behind the Property Graph model and are developing Property Graph-based data management solutions.

The third driver is a shift in the interest of analytics from merely reporting towards data-intensive science and discovery [6]. One major method among others in this discipline is network analysis, which puts the focal point of interest on the connectivity of entities. The toolbox of network analysis offers a rich set of algorithms and measures. These tools give incentives to consider the graph structure of data collections in a wide range of application fields, further increasing the demand for scalable graph data management solutions.

Today, graph data management has become a major topic in the database community, in research as well as industry. With more and more data available in graph-structured form, there is an increasing need to perform analytics on graph data. Descriptive multidimensional analytics is known since the 1980s [7] and part of the very fundamental analytics toolbox. Hence, multidimensional analytics on graphs is quickly becoming a hot topic in the literature and got tackled from various angles and settings. From the perspective of graph database management systems, a base requirement that comes with multidimensional analytics is to provide multidimensional graph query functionality. One essential question is how multidimensional graph queries can be formulated in a declarative graph query language. This question is important because the query language is one of the most crucial elements of any database management system (DBMS). It defines the functionality of a DBMS and how it is exposed to the user.

In this article, we take a structured and detailed look at the question of formulating declarative multidimensional graph queries. Thereby, the article focuses on query language definition and semantics. Query processing is not considered. We start by giving an introduction to graph data models in Sect. 2. Afterwards, Sect. 3 presents subgraph matching as the foundation of declarative graph queries. Section 4 considers graph transformations as the main concept to make graph queries composable, which is an important property particularly in analytics. Section 5 discusses declarative multidimensional graph queries based on the foundations that are laid before. Finally, Sect. 6 concludes the article.

2 Graph Data Models

There are multiple ways to define a graph. Let's start with the simplest one, the directed graph.

Definition 1 (Directed Graph). *A directed graph $G(V, E)$ consists of a set of vertices V and a set of edges $E \subseteq V \times V$. Two vertices $v, u \in V$ are adjacent if there exists an edge (v, u) or (u, v) in E. Let $\mathtt{src}(\cdot) : E \to V$ be a function yielding the source vertex of an edge and $\mathtt{trg}(\cdot) : E \to V$ be a function yielding the target vertex of an edge so that $e = (\mathtt{src}(e), \mathtt{trg}(e))$ holds.*

Also undirected graphs are possible where $(v, u) \in E \leftrightarrow (u, v) \in E$. However, we do not consider them any further here, since directed graphs can easily represent undirected ones by duplicating every edge to the other direction.

In addition to the basic building blocks—vertices and (directed) edges—there are more elements typically in use to enrich a graph data model. Figure 1 shows the most relevant elements. First, vertices and edges can be augmented with labels. Vertex labels are sometimes used as user-given vertex identities. In that case each vertex label has to be unique in the graph. Another use of vertex labels sees them as some arbitrary, not necessarily unique chunk of data associated with the vertex. The interpretation of this data is left to the user. Here, we consider vertex labels as typing information, which specifies what kind

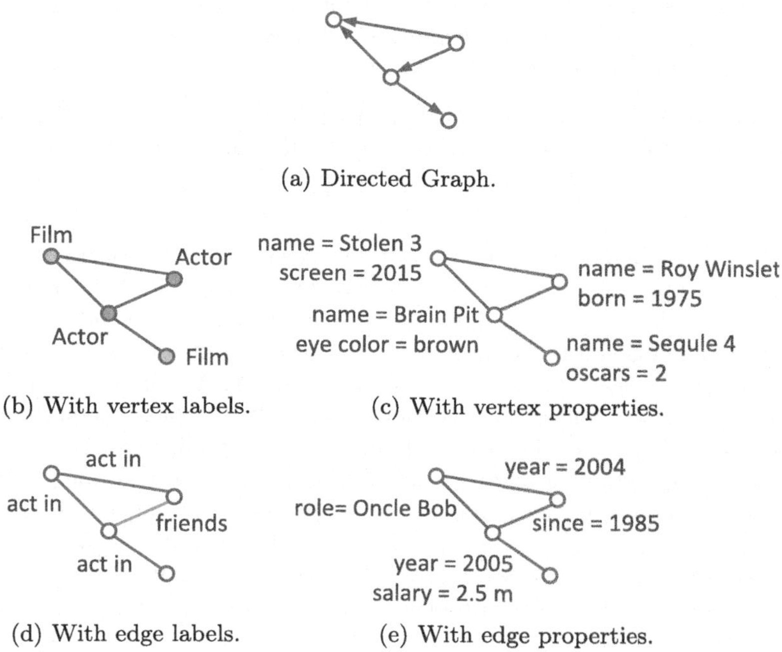

Fig. 1. Graph building blocks by example.

of entity, e.g. Film and Actor, is represented by a vertex. Edge labels are typically also used as typing information for the relationships between entities, e.g. act in and friends. Sometimes edge weights are referred to as edge labels. However, we assume edge labels to be typing information here. Second, vertices and edges can be augmented with properties, i.e. key–value pairs. Properties are a structured way to associate a graph with data, while the use of labels for that purpose is an unstructured way.

These additional building blocks can be very freely combined, resulting in a zoo of possible graph data models. Further differences between graph data models are whether loops (edges between one and the same vertex) are allowed and multigraphs (multiple edges between the same pair of vertices) are supported. Some graph models consider edges as objects with own identities. This is particularly handy in multigraphs.

Because of this diversity in graph models, most research papers and articles on graph data management and analytics provide a definition of the graph model used. Typically, the graph model is defined as simple as the considered problem allows to avoid unnecessary complexity. This is important and requires careful consideration particularly when the presented concepts and findings should be applied in another context which may use a different graph model.

For data management a plethora of graph data models has been proposed and studied over the years. Excellent overviews can be found in [8,9]. Most commonly used in the data management domain are labeled multigraphs and property graphs.

RDF data is an example of a labeled multigraph, more precisely an edge-labeled multigraph. Edge-labeled multigraphs have edge labels and allow arbitrarily many edges between a pair of vertices.

Definition 2 (Edge-labeled Multigraph). *An edge-labeled multigraph* $G(V, E, \rho, \Sigma, \lambda)$ *consists of a set of vertices* V*, a set of edges* E*, an incidence function* $\rho : E \rightarrow V \times V$*, and a labeling function* $\lambda : E \rightarrow \Sigma$ *that assigns a label to each edge.*

RDF stands for Resource Description Framework, a W3C-standardized data model to describe web resources [10]. It provides interoperability between applications that exchange machine-understandable information on the Web; it is the foundation of the semantic web. Essentially, RDF builds on two ideas. First, a data set is a collection of statements about resources. A statement consists of a subject, a predicate, and an object, hence also called triple. The subject is the resource the statement declares something about; the predicate is the property of the resource the statement declares; the object is the property value the resource has. Second, everything is a web resource—subjects, predicates, objects, and meta data, such as types—so that statements can be made about everything. The only exception are literal values such as numbers, strings, dates, etc. Only objects can be literals. Web resources are identified by Internationalized Resource Identifiers (IRIs) [11]—the internationalized variant of Uniform Resource Identifiers (URIs) [12]. A set of triples forms an RDF graph—an edge-labeled multigraph. Figure 2 shows an example RDF graph consisting of five triples.

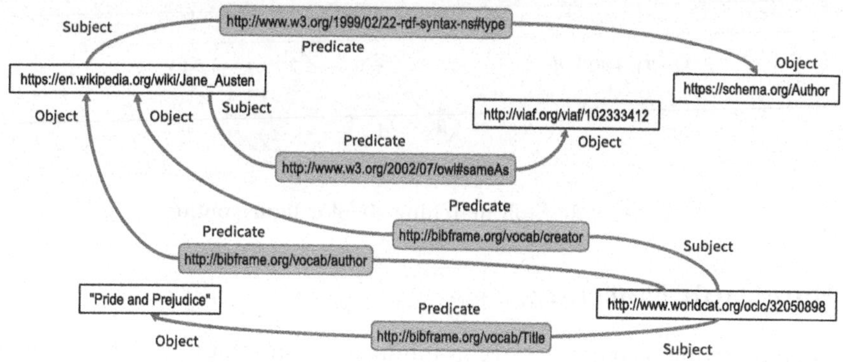

Fig. 2. Example of an RDF graph.

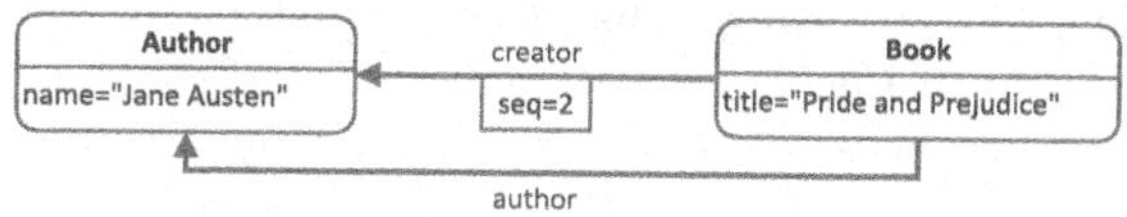

Fig. 3. Example of a property graph.

Property graphs [13] are also multigraphs but fully labeled and attributed, i.e. with edge and vertex properties. Hence, it can be seen as a graph of rich objects. Properties describe the features of these objects and labels are typically used to indicate object types.

Definition 3 (Property Graph). *A property graph* $G(V, E, \rho, \Sigma, \lambda, K, \mathcal{D}, \pi)$ *consists of a set of vertices* V*, a set of edges* E*, an incidence function* $\rho : E \rightarrow V \times V$*, a labeling function* $\lambda : V \cup E \rightarrow \Sigma$*, a set of property keys* K*, a domain of valid values* $\mathcal{D}$*, and a partial function* $\pi : (V \cup E) \times K \rightarrow \mathcal{D}$ *that assigns to a vertex or an edge and a property key a value if the vertex or the edge has the given property.*

Property graphs became popular with Neo4j[1] and TinkerPop[2] and are now used and implemented by many others, such as IBM Graph[3], Oracle [14], SAP [15,16], and Sparksee[4] [17,18]. Although there is a common understanding what property graphs are, there exists no standard so far and implementations differ in subtleties. Figure 3 shows an example property graph consisting of two vertices and two edges.

For clarity, definitions throughout this article are based on simpler graph models, such as directed graphs or labeled graphs. Only the elements necessary to define the essence of a concept are included in its definition. Which kind of graph model a definition is based on is either stated or obvious.

---

1 https://neo4j.com/.
2 http://tinkerpop.apache.org/.
3 https://ibm-graph-docs.ng.bluemix.net/.
4 http://www.sparsity-technologies.com/.

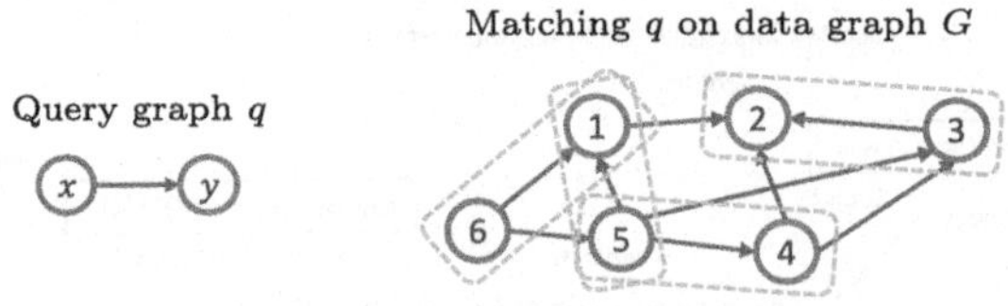

Fig. 4. Subgraph matching. (Color figure online)

3 Subgraph Matching

Subgraph matching is a declarative graph querying mechanism, where the query is given by a *query graph* q. The literature often refers to the query graph as a *pattern* and to subgraph matching as *pattern matching*, analogically. Intuitively, subgraph matching finds all subgraphs in *data graph* G that fit the given query graph q, as illustrated in Fig. 4.

In Fig. 4, the query asks for a vertex x with label Green having a directed edge to a vertex y with label Red. In the data graph, vertex 3 and vertex 2 and the directed edge between both form the subgraph $(V, E) = (\{3, 2\}, \{(3, 2)\})$ that fits this pattern. For that result vertex 3 corresponds to x and vertex 2 corresponds to y. Such a mapping between vertices in the query graph and vertices in the data graph, which describes one result of the query is often denoted as an *embedding*. Further intuitive results are $(\{6, 1\}, \{(6, 1)\})$, $(\{5, 1\}, \{(5, 1)\})$, and $(\{5, 4\}, \{(5, 4)\})$. For a more rigor and formal definition of subgraph matching let us first consider graph similarity.

3.1 Graph Similarity

Graph similarity considers the question whether two graphs are similar regarding their structure. For instance consider Fig. 5. Do we consider the structure of the graphs G and H to be similar, yes or no?

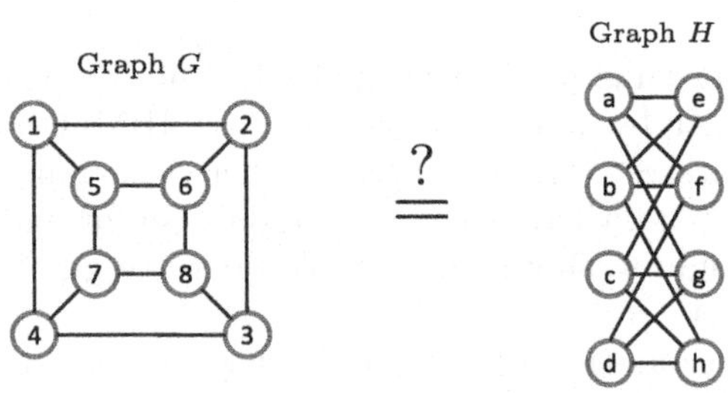

Fig. 5. Graph similarity.

Obviously the answer to that question is application-dependent. Consequently, numerous graph similarities for various applications are defined in literature. Most graph similarities are based on a relation σ that has to exist between the respective elements, typically the vertices, for two graphs have to

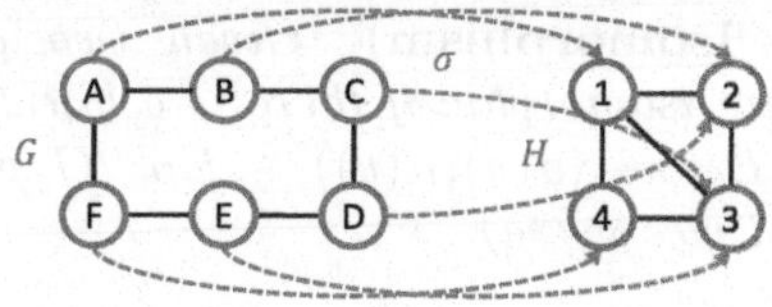

Fig. 6. Graph homomorphism.

be considered as similar. The various similarities differ in the additional conditions that have to hold for σ. Depending on the conditions, graph similarity semantics range from very lose similarities to very tight ones. The graph similarities most widely used in data management are *homomorphism* and *isomorphism*, which we will detail in the following.

Definition 4 (Graph Homomorphism). *Given two graphs* $G(V_G, E_G)$ *and* $H(V_H, E_H), G$ *is homomorphic to* H *if there is a surjective*[5] *function* $\sigma : V_G \rightarrow V_H$ *such that* G *preserves adjacency of* H*, i.e.* $(u, v) \in E_G \rightarrow (\sigma(u), \sigma(v)) \in E_H$.

Figure 6 illustrates the definition of graph homomorphism with a small example of graphs G and H. As can be seen, graph homomorphism requires that there exists a relation σ from V_G to V_H which is a surjective function and that the vertices in G have the same adjacency as their counterparts in H according to σ. In the example, σ relates E to 3 and D to 2. Since there is an undirected edge between E and D in G, there has to be also an indirected edge between 3 and 2 in H. The same has to hold for all other edges in G, which is the case so that G is homomorphic to H.

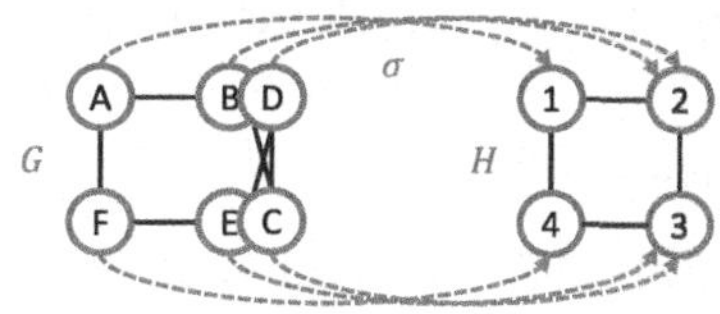

Fig. 7. Graph homomorphism intuition.

The intuition behind graph homomorphism is illustrated in Fig. 7. Here G is drawn slightly differently. Thinking visually, we can say that G is homomorphic to H if we can move the vertices of G around—including putting them on top of each other—in such a way that G appear identical to H. Considering the definition as well as the intuition, it becomes clear that graph homomorphism is an asymmetric similarity. In the example, G is homomorphic to H but not vice versa.

In contrast, the more strict graph isomorphism provides a symmetric similarity.

[5] Left-total, right-total, and right-unique.

Definition 5 (Graph Isomorphism). *Given two graphs $G(V_G, E_G)$ and $H(V_H, E_H)$, G and H are isomorphic if there is a bijective*[6] *function $\sigma : V_G \rightarrow V_H$ such that $(u, v) \in E_G \leftrightarrow (\sigma(u), \sigma(v)) \in E_H$ (G preserves adjacency and non-adjacency of H and vice versa).*

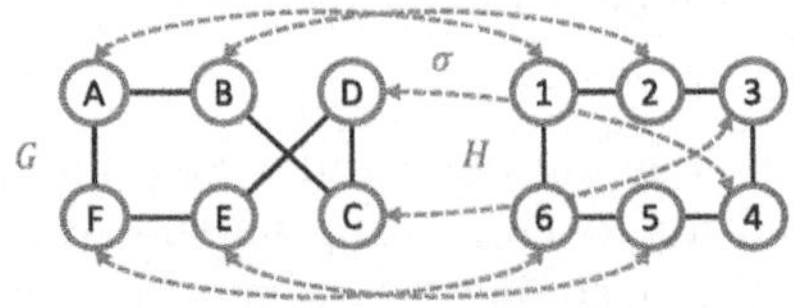

Fig. 8. Graph isomorphism.

In contrast to homomorphism, graph isomorphism requires that σ is a bijective function and that vertices of both, G and H have the same adjacency as well as non-adjacency as their counterparts in H and G, respectively. Figure 8 shows two isomorphic graphs G and H. In this example, σ relates C with 3, D with 4, and E with 5. Since there is an undirected edge between C and D and between D and E in G, also an undirected edge between 3 and 4 and between 4 and 5 in H has to exist. Likewise, since there is no edge between C and E in G, there must not be an edge between 3 and 5 in H. The same has to hold for all pairs of vertices in both directions, which is the case so that G and H are isomorphic. The intuition behind graph isomorphism is similar to homomorphism with a crucial exception. Thinking visually, we can say G and H are isomorphic if we can move the vertices of G or H around—*without* putting them on top of each other—in such a way that both look identical.

3.2 Matching Semantics

Graph similarity can be utilized in subgraph matching by turning it into a selection criteria. While the graph similarity itself asks for a σ with certain constraints to decide similarity, subgraph matching asks for minimal subgraphs—no extra data—of a data graph for which such a σ exists so that similarity is given.

Definition 6 (Subgraph Homomorphism). *Given a query graph $Q(V_Q, E_Q)$ and a data graph $G(V_G, E_G)$, a graph $R(V_R, E_R)$ is a result for Q on G if*

- *$V_R \subseteq V_G$ and $E_R \subseteq E_G$ (R is a subgraph of G); and*
- *there is a surjective function $\sigma : V_Q \rightarrow V_R$ such that $(u, v) \in E_Q \leftrightarrow (\sigma(u), \sigma(v)) \in E_R$ (Q preserves adjacency of R with no extra edges in R).*

Figure 9 illustrates subgraph homomorphism. The figure shows a query graph Q, an excerpt of a data graph G, and one subgraph R matching Q under homomorphism. As can be seen, R is a minimal subgraph to which Q is homomorphic.

[6] Left-total, left-unique, right-total, and right-unique.

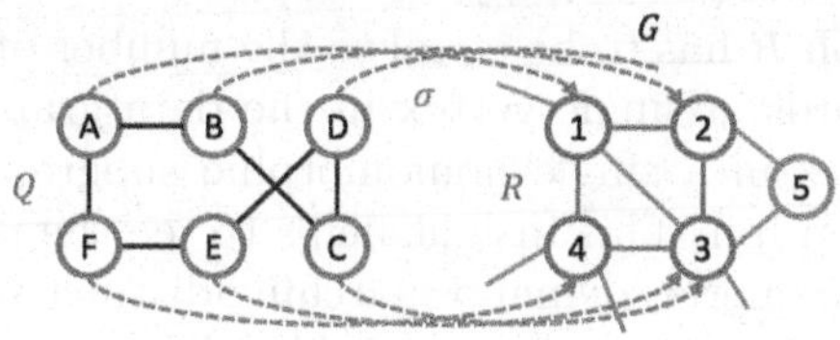

Fig. 9. Subgraph homomorphism.

If we would include edge $(1, 3)$ in R, Q would remain homomorphic to R according to Definition 4. However, R would contain additional data the query does not ask for. Definition 6 prevents such extra data with equivalence—instead of implication—in the adjacency preservation condition. Similar to graph homomorphism, subgraph homomorphism allows to map a different query vertices (e.g. B and D) to the same vertex (e.g. 2) in the data graph. In consequence, the matched subgraph may have less vertices than the query graph.

In labeled graphs and property graphs, subgraph matching additionally demands, next to the structural similarity between the query graph and resulting subgraphs, a similarity between the matched vertices and edges and their counterparts in the query graphs. The definition can be easily extended to capture that by additionally requiring that $\forall (v_Q, v_R) \in \sigma, v_Q \sim v_R$ (vertex label and properties match) and $\forall (u, v) \in E_Q, (u, v) \sim (\sigma(u), \sigma(v))$ (edge label and properties match) holds for σ.

Similar to subgraph homomorphism, graph isomorphism can be utilized for matching, cf. [19].

Definition 7 (Subgraph Isomorphism). *Given a query graph $Q(V_Q, E_Q)$ and a data graph $G(V_G, E_G)$, a graph $R(V_R, E_R)$ is a result for Q on G if*

- $V_R \subseteq V_G$ *and* $E_R \subseteq E_G$ *(R is a subgraph of G); and*
- *there is a bijective function* $\sigma : V_Q \to V_R$ *such that* $(u, v) \in E_Q \leftrightarrow (\sigma(u), \sigma(v)) \in E_R$ *(Q preserves adjacency of R with no extra edges in R).*

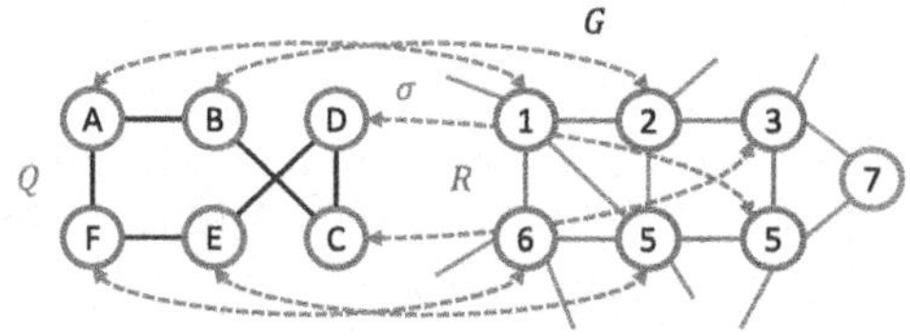

Fig. 10. Subgraph isomorphism.

The matching semantics can be tied to subgraph isomorphism by requiring σ to be a bijective function. As can be seen by comparing Figs. 9 and 10, the effect of isomorphism in contrast to homomorphism is that the number of vertices in the matching subgraph R has to be equal to the number of vertices in the query graph Q. In other words, a single vertex in the data graph can be matched by multiple query vertices in a single homomorphic subgraph but only once in a single isomorphic subgraph. This insight helps to see the usefulness of subgraph homomorphism. Often queries demand matching distinct vertices within a single subgraph. For instance, consider a query asking in a social network for pairs of friends and the respective city each of the friends lives in, as shown in Fig. 11. In the example, subgraph isomorphism finds only friends living in different cities, which is unlikely the intention of the query author. Subgraph homomorphism also finds the friends Mary and Anne both living in Berlin.

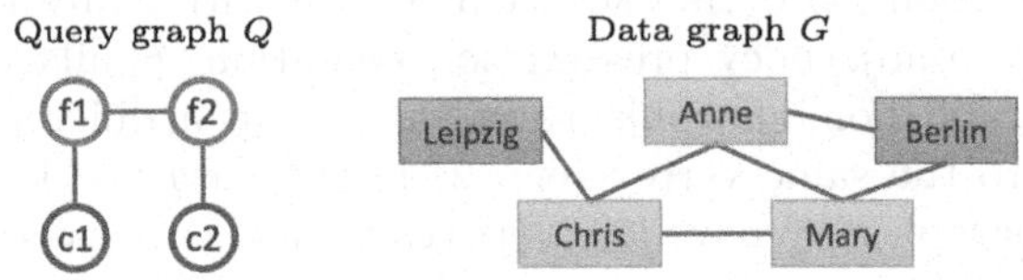

Fig. 11. Example illustrating the need for subgraph homomorphism.

The query language Cypher of the popular graph data management system Neo4j employs a slightly different matching semantics. Assume a movie graph database that connects movies, actors, directors, and reviewers as illustrated in Fig. 12. With the Cypher query

```
MATCH (a1)-[:ACTED_IN]-(m)-[:ACTED_IN]-(a2) WHERE a1=a2 RETURN a1,a2
```

we can ask for persons that acted in the same movie (`MATCH` clause) and filter for pairs of identical persons (`WHERE` clause). In Neo4j, this query returns an empty result set. This suggests Neo4j employs isomorphism as matching semantics, since query vertices `a1` and `a2` are not allowed to match the same vertex in the data graph—homomorphism would allow such results. Assume we ask for reviewer pairs, with the following query

```
MATCH (m1)<-[:REVIEWED]-(r1)-[:FOLLOWS]->(r2)-[:REVIEWED]->(m2)
WHERE m1=m2 RETURN m1,r1,r2,m2
```

where both reviewers (`r1` and `r2`) review a movie (`m1` and `m2`, respectively) and one reviewer follows the other (e.g. on Twitter). Additionally, we filter for results where the reviewers review the same movie (`m1=m2`). Again, the result set should be empty if Neo4j employs isomorphism for the same reasoning as with the first query. However, the query returns a non-empty result set allowing equal bindings for query vertices `m1` and `m2`. This suggests Neo4j employs homomorphism, obviously contradictory with the first finding.

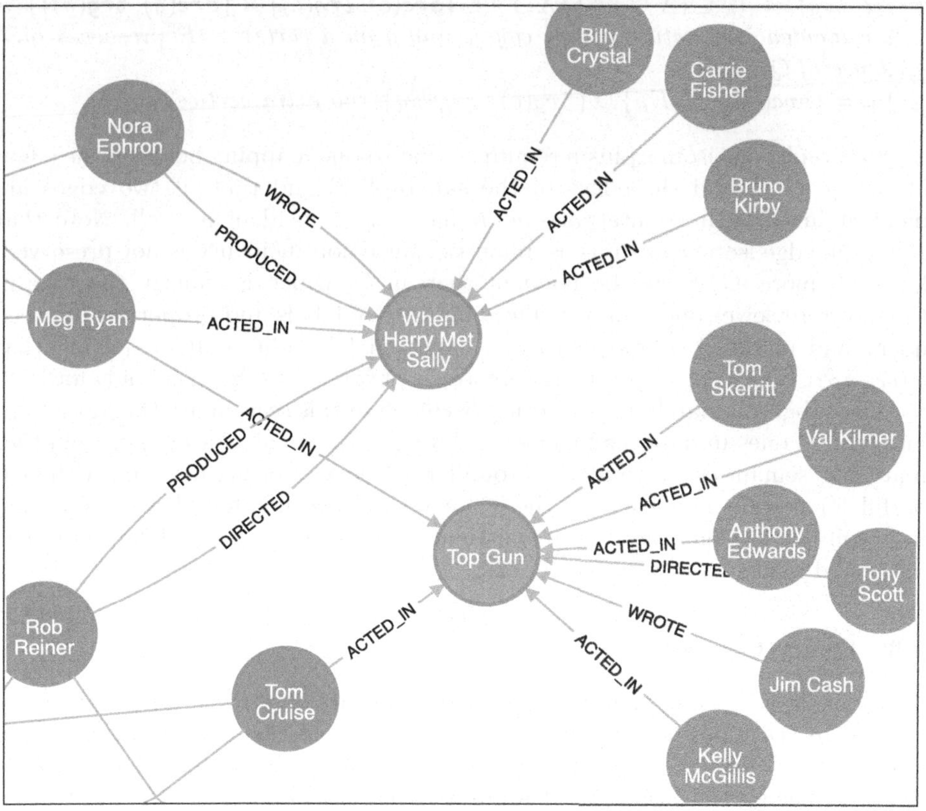

Fig. 12. Excerpt of the movie graph.

The Cypher Manual indicates that Cypher's matching semantics is based on edges instead of vertices[7]. Informally, "Cypher makes sure to not return matches where [two query edges] point to the same [sub]graph [edge]", as described in the manual. Formally, Cypher's matching semantics can be defined as subgraph edge isomorphism.

Definition 8 (Subgraph Edge Isomorphism). *Given a query graph $Q(V_Q, E_Q)$ and a data graph $G(V_G, E_G)$, a graph $R(V_R, E_R)$ is a result for Q on G if*

- $V_R \subseteq V_G$ *and* $E_R \subseteq E_G$ *(R is a subgraph of G);*
- *there is a bijective function* $\sigma^E : E_Q \rightarrow E_R$ *and a surjective function* $\sigma^V : V_Q \rightarrow V_R$ *such that* $(e, v, g) \in E_Q \times V_Q \times E_Q \wedge I(e, v, g) \rightarrow (\sigma^E(e), \sigma^V(v), \sigma^E(g)) \in E_R \times V_R \times E_R \wedge I_G(\sigma^E(e), \sigma^V(v), \sigma^E(g))$ *where*

[7] http://neo4j.com/docs/developer-manual/current/cypher/introduction/uniqueness/.

$I(e, v, g) = (v \in (X \cup Y) \mid (X, Y) \in \{src(e), trg(e)\} \times \{src(g), trg(g)\})$ *is the incidence signature of two edge e and g on a vertex v (R preserves incidence of Q); and*

- $V_R = \{src(e) \mid e \in E_R\} \cup \{trg(e) \mid e \in E_R\}$ *(no extra vertices in R).*

Subgraph edge isomorphism requires a one-to-one mapping between the edges of the query Q and the edges of the subgraph R and that, if two edges are incident in Q, their counterparts in R have to be incident as well. Note that subgraph edge isomorphism is asymmetric, since non-incidence is not preserved, i.e. in R more edges can be incident than in Q, which is exactly the case in the movie reviewer query above. The similarity of labels and properties between query and matched subgraph can be captured by additionally requiring that $\forall (e_Q, e_R) \in \sigma, (e_Q \sim e_R \wedge \texttt{src}(e_Q) \sim \texttt{src}(e_R) \wedge \texttt{trg}(e_Q) \sim \texttt{trg}(e_R))$ holds for σ.

The subgraph matching semantics discussed in this section are the most commonly used ones in data management. However, there exists a plethora of other matching semantics particularly around the concept of simulation. Although useful for certain applications, these semantics have not found wide adoption in graph data management systems. Hence, we do not discuss them here. The interested reader may refer to [20–22].

3.3 Query Classes

The idea of subgraph matching forms the nucleus of a range of graph query classes. In the following, we discuss the most important query classes, namely *Conjunctive Queries*, *Regular Path Queries* and *Conjunctive Regular Path Queries*. The classification and definitions given here follow [23]; the illustrating example is also inspired by [23]. All definitions are based on an edge-labeled graph. However, the query classes are rather general and their definition can be easily adapted to fit other kinds of graphs such as property graphs.

Conjunctive Queries (CQs) allow querying subgraph patterns. A query is given as a set of edge predicates. Each edge predicate consists of a pair of vertex variables and an edge label. The set of edge predicates forms a subgraph pattern.

Definition 9 (Conjunctive Query). *A conjunctive query Q is an expression*

$$(z_1, \ldots, z_n) \leftarrow \bigwedge_{1 \leq i \leq m} (x_i, a_i, y_i)$$

where each variable $x_i \in X$ *and* $y_i \in Y$ *is a vertex variable or a constant from* V*, each* $a_i \in \Sigma$ *is an edge label, and each* z_i *is a vertex variable from* $X \cup Y$*.*

A set of variable bindings is a valid answer iff all predicates hold on the data graph.

Definition 10 (Conjunctive Query Semantics). *Let $\sigma : X \cup Y \to V$ be a specific selection of variable bindings, i.e. a mapping*[8] *to the vertices of the labeled graph $G(V, E, \Sigma, \lambda)$. Say relation $(G, \sigma) \models Q$ holds iff $(\sigma(x_i), \sigma(y_i)) \in E$ and $\lambda((\sigma(x_i), \sigma(y_i))) = a_i$ for $1 \leq i \leq m$, i.e. σ maps the query pattern to a valid subgraphs in G. The query result $Q(G)$ is the set of tuples $(\sigma(z_1), \ldots, \sigma(z_n))$ such that $(G, \sigma) \models Q$.*

As a set of variable bindings is a surjective function—every variable has to be bound to exactly one vertex, different variables can be bound to the same vertex—conjunctive queries find subgraphs to which the subgraph pattern is homomorphic. However, by changing the requirements for σ the definition can easily be adapted to other matching semantics such as isomorphism. Hence, query classes—although typically defined with subgraph homomorphism—can be considered as orthogonal to the matching semantics.

Figure 13 shows an example graph, a conjunctive query, and the resulting variable bindings. The query asks for all x who have won the Nobel and the Booker prize and were born in y. For the second result [x = Gordimer, y = South Africa] the figure also shows the corresponding subgraph.

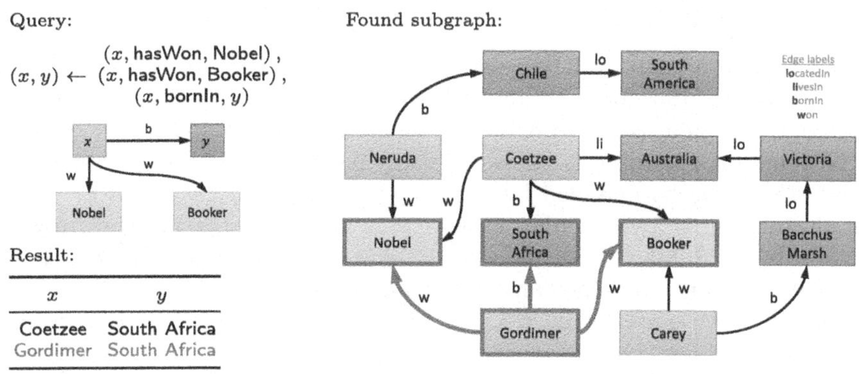

Fig. 13. Example of a conjunctive query.

Note that CQs allow to project the found variable bindings to a subset of variables. For instance the example query $(x, y) \leftarrow (x, \mathsf{hasWon}, \mathsf{Nobel}), \ldots$ in Fig. 13 projects to variables x and y. If we change the query to $(y) \leftarrow (x, \mathsf{hasWon}, \mathsf{Nobel}), \ldots$, it would project to variable y only so that it returns only countries of birth of the prize winners but not the prize winners themselves. Since CQs have set semantics for the result set, the only result in that case would be [South Africa].

Queries that additionally allow uniting the result sets from different conjunctive subqueries are called *Union Conjunctive Queries* (UCQs).

[8] A mapping is a surjective function.

Regular Path Queries (RPQs) allow querying reachability. A query is given as a single path predicate consisting of a pair of vertex variables and a path expression. The path expression is a regular expression over edge labels.

Definition 11 (Regular Path Query). *A regular path query Q is an expression*

$$(x, y) \leftarrow (x, r, y)$$

where variables x and y are vertex variables or a constant from V and $r \in \Sigma^*$ is a regular expression over the alphabet of edge labels Σ.

A pair of variable bindings is a valid answer iff the respective vertices are connected in the data graph by a path conforming to the path expression.

Definition 12 (Regular Path Query Semantics). *A path p between v_0 and v_m in a labeled graph $G(V, E, \Sigma, \lambda)$ is a sequence $v_0 a_0 v_1 a_1 \ldots v_{m-1} a_{m-1} v_m$, with $v_i \in V$ for $i \in [0, m]$, $a_i \in \Sigma$, $(v_i, v_{i+1}) \in E$, and $\lambda((v_i, v_{i+1})) = a_i$ for $i \in [0, m)$. Let $\lambda(p) \in \Sigma^*$ be the label of the path, with $\lambda(p) = a_0 a_1 \ldots a_{m-1}$. Let $L(r)$ be the language denoted by the regular expression r, i.e. the set all of all possible path labels denoted by r. Path p satisfies r if $\lambda(p) \in L(r)$, i.e. the p's label satisfies the regular expression r. The query result $Q(G)$ is the set of all pairs of vertices (x, y) in G such that there is a path from x to y that satisfies r.*

Regular expressions for RPQs allow edges labels as literals and concatenation (denoted $\cdot$ or /), alternation ($|$), and Kleene star (*) as expressions. The concatenation of two regular path expressions r and s matches any path that consists of any path match by r directly followed by any path matched by s. More formally, if $m(r)$ are all paths matched by r, $m(r \cdot s) = \{p_r p_s \mid (p_r, p_s) \in m(r) \times m(s)\}$. The alternation of two regular path expressions r and s matches any path matched by r or by s, i.e. $m(r \mid s) = m(r) \cup m(s)$. The Kleene star on a regular expression r matches any finite number (including zero) of concatenations of any path matched by r, i.e. $r^* = r^{[0,\infty]}$, where $m(r^0) = \epsilon$ (zero-length path) and $r^{i+1} = r^i \cdot r$ and $r^{[l,u]} = \bigcup_{i \in [l,u]} r^i$ with $m(r^{[l,u]}) = \bigcup_{i \in [l,u]} m(r^i)$. The zero-length path is defined to connect every vertex with itself. Hence, each pair (v, v) with vertex $v \in V$ is part of any RPQ with a regular expression of the form r^*. Specific regular expression implementations may allow additional convenience expressions. $r?$ matches zero or one of any that is matched by r, i.e. $r? = \epsilon \mid r = r^{[0,1]}$. r^+ matches any non-zero number of concatenations of any path matched by r, i.e. $r^+ = r^{[1,\infty]}$ and $m(r^+) = m(r^*) \setminus \{\epsilon\}$.

Figure 14 shows an example of a regular path query. The query asks for all authors x and the location y of their place of birth or living place. The connection between an author x and an location y is specified by the query with the regular path expression $(\mathsf{b}|\mathsf{li}) \cdot \mathsf{lo}^*$. Matching paths have to start with a b edge or a li edge followed by zero or more lo edges. Labels of paths that would match are (b), (li), $(\mathsf{b}, \mathsf{lo})$, $(\mathsf{li}, \mathsf{lo})$, $(\mathsf{b}, \mathsf{lo}, \mathsf{lo})$, $(\mathsf{li}, \mathsf{lo}, \mathsf{lo})$, etc. Hence, the query finds all x and y connected by path with such a label. For the first result $[x = \mathsf{Carey}, y = \mathsf{Australia}]$ the figure also shows the corresponding path with label $(\mathsf{b}, \mathsf{lo}, \mathsf{lo})$.

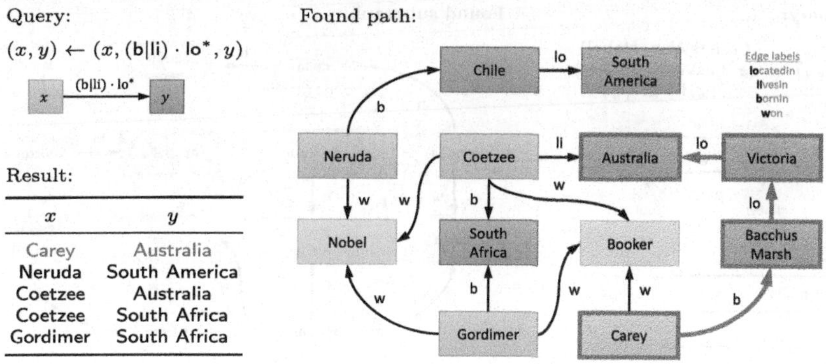

Fig. 14. Example of a regular path query.

In directed graphs it is often necessary to search for reachability along reverse edge direction. For instance in the example graph, we may want to ask for pairs that have won the same prize. RPQs can be extended by a regular expression r^- (or $\hat{r}$) that matches any path that has a reverse path matched by r. Queries supporting reverse direction matching are called *2-way Regular Path Queries* (2RPQs). RPQs can also be extended to property graphs. It is straightforward to extend regular expressions from simple label equivalence predicates to more general edge predicates—predicates that can be evaluated isolatedly on an individual edge. Remembering data along the path, i.e. allowing predicates across multiple edges in the path, is not straightforward, though. A query would require a predicate across multiple edges if, for instance, it needs to check that the first edge and the last edge of the path have the same value on a certain property (e.g. the same creation date) without restricting the property to one particular value. Such extensions are discussed in [24–27].

Conjunctive Regular Path Queries (CRPQs) combine subgraph pattern querying with reachability querying. A query is given as a set of path predicates. Each path predicate consists of a pair of vertex variables and a regular expression over edge labels.

Definition 13 (Conjunctive Regular Path Query). *A conjunctive regular path query Q is an expression*

$$(z_1, \ldots, z_n) \leftarrow \bigwedge_{1 \leq i \leq m} (x_i, r_i, y_i)$$

where each variable $x_i \in X$ and $y_i \in Y$ is a vertex variable or a constant from V, each $r_i \in \Sigma^*$ is a regular expression over the alphabet of edge labels Σ, and each z_i is a vertex variable from $X \cup Y$.

A set of variable bindings is a valid answer iff all path predicates hold on the data graph.

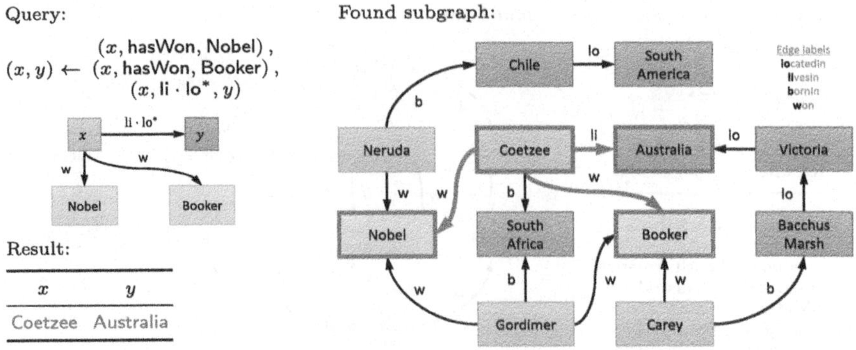

Fig. 15. Example of a conjunctive regular path query.

Definition 14 (Conjunctive Regular Path Query Semantics). *Let* $\sigma : X \cup Y \rightarrow V$ *be a specific selection of variable bindings, i.e. a mapping to the vertices of the labeled graph* $G(V, E, \Sigma, \lambda)$. *Say relation* $(G, \sigma) \models Q$ *holds iff, for* $1 \leq i \leq m$ *there exists a path* p_i *in* G *from* $\sigma(x_i)$ *to* $\sigma(y_i)$ *such that* $\lambda(p) \in L(r_i)$. *The query result* $Q(G)$ *is the set of tuples* $(\sigma(z_1), \ldots, \sigma(z_n))$ *such that* $(G, \sigma) \models Q$.

Figure 15 shows an example of a conjunctive regular path query. The query asks for all x who have won the Nobel and the Booker prize where the living place is located (y). For the only result [$x =$ Coetzee, $y =$ Australia] the figure also shows the corresponding subgraph.

With unions and reverse edge matching added, queries are called *Union Conjunctive 2-way Regular Path Queries* (UC2RPQs). Further, the literature discusses a plethora of other query classes, mainly in the search for query classes that have low complexities for query evaluation, containment, and other features while being as powerful as possible, for instance [28–30].

3.4 Query Languages

Various graph query languages based on subgraph matching have already been implemented. A few should be mentioned here.

SPARQL. The query language for RDF graphs, SPARQL is a long existing W3C standard and was updated to version 1.1 in 2013 [31,32]. Minimally, a SPARQL query consists of a `SELECT` clause and a `WHERE` clause. The optional `PREFIX` clause allows abbreviating IRIs of the RDF data with namespaces. The query from Fig. 15 can be expressed in SPARQL as:

```
PREFIX my: <http://myrdf.com/>
SELECT ?x ?y
WHERE { ?x my:hasWon                  my:nobel  .
        ?x my:hasWon                  my:booker .
        ?x my:livesIn/my:locatedIn* ?z .          }
```

The `WHERE` clause takes a basic graph pattern, which is (simplified) a set of RDF triples with variables, so called triple patterns. Variables, denoted with question mark, can be listed in the `SELECT` clause to form the result set. SPARQL employs essentially subgraph homomorphism as matching semantics and supports all of the discussed query types up to UC2RPQs. Regular path expressions can be denoted as so called *property paths* in the predicate position of a triple, as shown in the example. Additionally, SPARQL allows additional filtering including filtering for non-existence of triples, optional matches, and set difference. SPARQL is widely implemented, for instance in Virtuoso[9], Blazegraph[10], AllegroGraph[11], Apache Jena[12], RDF4J[13], and RDFox [33].

Neo4j Cypher. The graph database management system Neo4j provides the query language Cypher to query property graphs. A minimal Cypher query consists of a `MATCH` clause and a `RETURN` clause. The query from Fig. 15 can be expressed in Cypher as:

```
MATCH (x)-[:HAS_WON]->(:Prize {name='Nobel'}),
      (x)-[:HAS_WON]->(:Prize {name='Booker'}),
      (x)-[:LIVES_IN]->()-[:LOCATED_IN*]->(y)
WITH DISTINCT x, y
RETURN x.name, y.name
```

The assumption here is that the prizes are presented as vertices with the label Prize and the property name giving the name of the prizes. Persons and location are also assumed to have the property name. As has been discussed in Sect. 3.2 already, Cypher employs a rather special matching semantics that can be described as subgraph edge isomorphism. Furthermore, the support for RPQs is rather limited. Within a pattern, Cypher only allows the Kleene star with optional lower and upper bounds. That allows expressing simple regular paths as in the query from Fig. 15. However, Cypher's Kleene star is limited to a single edge label (including a wildcard matching all edge labels). Kleene star over a concatenation, alternation, or another Kleene star expression cannot be expressed in Cypher. On top of this functional limitation, there is a difference in semantics. More precisely, Cypher's Kleene star queries unique paths instead of reachability. As there are typically multiple paths between a pair of vertices, this specialty requires duplicate elimination to get reachability results. Duplicate elimination can be easily done with the `WITH` clause and the keyword `DISTINCT`. Since Cypher queries for distinct paths and there can be terribly many in a graph [34], using the pure Kleene star is in most cases not advisable; queries simply never come back. To prevent such unfortunate behavior the Kleene

[9] http://virtuoso.openlinksw.com/.
[10] https://www.blazegraph.com/.
[11] http://franz.com/agraph/allegrograph/.
[12] https://jena.apache.org/.
[13] http://rdf4j.org/.

star can be restricted to query only shortest paths with `(x)-[:LIVES_IN]->(z), shortestPath((z)-[:LOCATED_IN*]->(y))`.

PGQL. PGQL [14] is a property graph query language developed by Oracle. Syntactically, PGQL borrows its main structure from SQL and the graph pattern syntax from Cypher. A minimal PGQL query has the well known `SELECT-FROM-WHERE` structure. The `FROM` clause names the graph that should be queried; the `WHERE` clause specifies the query pattern; the `SELECT` clause defines the structure of the result. The query from Fig. 15 can be expressed in PGQL as:

```
SELECT x.name, y.name
FROM literatureGraph
WHERE (x)-[:hasWon]->(nob WITH name='Nobel'),
      (x)-[:hasWon]->(bok WITH name='Booker'),
      (x)-[:livesIn]->(z),
      (z)-/locatedIn*/->(y)
```

PGQL has subgraph isomorphism as default matching semantics. But subgraph homomorphism can be used specifying `WHERE ALL` instead of `WHERE`. Regular paths are expressed with the so called path query connectors `-/` and `/->`, as shown in the query. As a specialty, PGQL allows to declare named regular path patterns with the `PATH` clause. Named regular path patterns can be used in the query pattern or the definition of other named regular path patterns.

Gremlin. The Graph traversal language Gremlin [35] is designed for property graphs, developed by Apache TinkerPop, and implemented by various other systems, e.g. DataStax Enterprise Graph[14], IBM Graph, OrientDB[15], Stardog[16], and Titan Distributed Graph Database[17]. A Gremlin query basically describes which steps a traverser should take on a graph. In that sense, Gremlin is primarily not a subgraph matching based language. Nevertheless, one of the various kinds of traversal steps Gremlin offers is a so called match step. The query from Fig. 15 can be expressed in Gremlin as:

```
g.V().match(
        __.as('x').out('hasWon').has('name', 'Nobel'),
        __.as('x').out('hasWon').has('name', 'Booker'),
        __.as('x').out('livesIn').as('z'),
        __.as('z').repeat(__.out('livesIn').dedup())
                  .emit().dedup().as('y'))
      .select('x','y').by('name')
```

[14] http://datastax.com/products/datastax-enterprise-graph.
[15] http://orientdb.com/.
[16] http://stardog.com/.
[17] http://titan.thinkaurelius.com/.

The match step takes a collection of traversal fragments that conjunctively define the subgraph pattern and matches the pattern with subgraph homomorphism. Regular path reachability can be queried as well. However, like the pattern, the regular expression has to be coded in traversal steps, such as `or`, `optional`, or `repeat`. Particularly, expressing Kleene star becomes very indirect because the deduplication of intermediate vertices has to be specified explicitly with `.dedup()` as in the example.

4 Graph Transformation

All the query classes we discussed so far are one-step queries. Although very useful for OLTP-like queries that interact within base data granularity, this type of queries is very limited regarding information aggregation. However, information aggregation and condensation is the main mean to increase the business value of data. While data is typically collected on a fine grained level, the analytical interest is of higher abstraction and aggregated nature. For instance in social graphs like Facebook's, the analytical interest is not in a particular subgraph of friends or likes, but in aggregated information, e.g. which users get their posts most liked or most retweeted and, hence, are strong influencers. Analytical processes typically preform stepwise aggregation and condensation of fine granular base data into information of high abstraction. In these processes aggregated information is often recombined with base data and aggregated again. For instance, the strong influencers are checked regarding the products they talk about to find the products or the product groups that the influencer find most interesting. Next, this influence information is drilled down to see whether product posts get the same influence spread as the influencer's other posts to find the influencers that are valuable for viral marketing. To support such stepwise aggregation in a declarative query language, the query language has to be composable and has to provide grouping and aggregation functionalities. In the following, we discuss how the graph matching queries can be extended towards composability and aggregation.

4.1 Composability

Composability of a query language is a feature that allows to formulate queries on the result of other queries of that language. In other words the query language is nestable, one query can be subquery of another. Composability (or nestability) is an important principle for the design of simple but powerful programming languages in general and query languages in particular [36]. What composability brings to a language is at best illustrated with LEGO pieces. The LEGO system allows to interlock pieces of different variety. Essentially, LEGO can be seen as a formal language with all types of LEGO pieces being the alphabet and the interlocking system being the grammar. The LEGO language is composable. Combined pieces practically become a new individual piece that can be interlocked and recombined with each other and all other pieces. Even with its very simple and easy to understand base elements, its composability makes the LEGO language very powerful and capable of expressing an uncountable variety [37]. As the same holds for query languages, composability is an essential feature for simple but powerful query languages.

The subgraph matching queries discussed in Sect. 3 lack composability. All these queries operate on a graph but produce a table, i.e. a set of tuples, as result. Consequently, a query result cannot be input to other queries of the same language. To have a query language composable over graphs, a query has to return a graph.

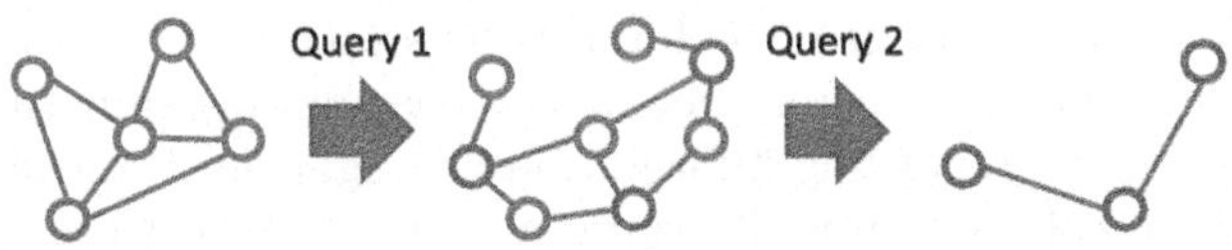

Fig. 16. Graph queries with a constructive element. (Color figure online)

A naïve way of making subgraph matching composable is to redefine it to return the matched subgraphs instead of a table of variable bindings. However, subgraph matching merely filters data. Composing multiple filters simply creates the conjunction of these filters. Even the simplest type of matching queries—conjunctive queries—already include conjunction. Hence, this naïve composition of matching queries does not increase the query language's power.

Unleashing the real power of composability requires a constructive element. Hence, a composable query must be able to return a graph different from subgraphs initially matched. As an example, consider the simple example in Fig. 16. The first query matches all triangles (three vertices fully connected among each other) of blue vertices and constructs red vertices. Essentially, each red vertex represents a matched triangle. Additionally, the first query connects each red vertex with each of the blue vertices that belong to the matched triangle a red vertex represents. The graph resulting from this first query is different from the subgraphs matched by the query. In fact, the resulting graph does not contain the initially matched triangles anymore but representatives of them. Such a graph can be used by a second query, which matches pairs of red vertices that are connected to the same blue vertex. As result the second query connects the matched pairs of red vertices with an edge. Like the first query, the second query returns a graph different from the subgraphs it matched. Essentially, the result graph of the second query shows how triangles are connected in the original graph. Such a composed chain of queries is not possible with pure matching queries as they are not able to construct new graphs from the matched subgraphs.

Constructing new graphs from the matched subgraphs is exactly what a graph transformation does. It combines subgraph matching with graph construction as illustrated in Fig. 17. Input to subgraph matching is a graph. Any of the matching queries discussed in Sect. 3 can serve the matching part. In any case,

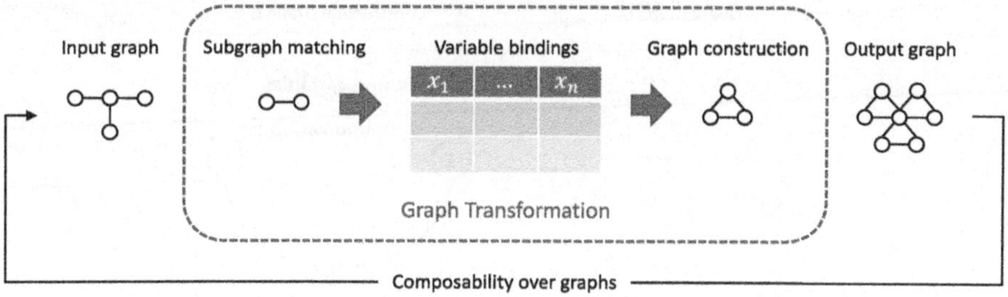

Fig. 17. Graph transformation.

the matching produces a table of variable bindings. The columns of such a table are vertex variables in the query pattern and the tuples are sets of variable–vertex bindings (embeddings) that constitute a valid result to the query. This table of variable bindings serves as input to graph construction. Output of the graph construction is a graph, so that graph transformations are composable over graphs.

Definition 15 (Graph Transformation). *A graph transformation $\Gamma_{Q,C} : \mathcal{G} \rightarrow \mathcal{G}$ is a function based on query graph $Q(V_Q, E_Q)$ and a construction graph $C(V_C, E_C)$ that maps a graph G to a graph G'. Given a subgraph matching procedure m and graph construction c, $\Gamma_{Q,C}(G) = c_C(m_Q(G)) = G'$.*

Like subgraph matching, graph construction is based on graph pattern. Each vertex in the construction graph has a variable assigned.

Definition 16 (Construction Graph). *A construction graph $C(V_C, E_C, \sigma_C)$ for a table T of variable bindings is a graph where $\sigma_C : V_C \rightarrow \mathcal{X}$ is an injective function assigning a variable to all vertices in C. A vertex $v \in V_C$ is either bound if its variable $\sigma_C(v)$ is in the schema of T ($\sigma_C(v) \in S(T)$) or unbound if $\sigma_C(v) \notin S(T)$.*

Essentially, the graph construction instantiates the construction graph with the variable bindings of every tuple in the bindings table. Variables in the construction graph which have a binding in the tuple are instantiated with the vertex in the table. Unbound variables in the construction graph are instantiated with a new vertex—new for every tuple. All instantiations form the resulting graph.

Definition 17 (Graph Construction). *Given a table T of variable bindings in a graph $G(V, E)$ and a construction graph $C(V_C, E_C, \sigma_C)$. Let $\pi : T \times V_C \rightarrow \mathcal{V}$ be a function that maps a combination of binding tuple and construction graph vertex to a vertex out of all possible vertices $\mathcal{V}$, so that*

$$\pi(t, v) = \begin{cases} t[\sigma_C(v)], & if\ \sigma_C(v) \in S(T) \\ u_t^v, & if\ \sigma_C(v) \notin S(T) \end{cases}$$

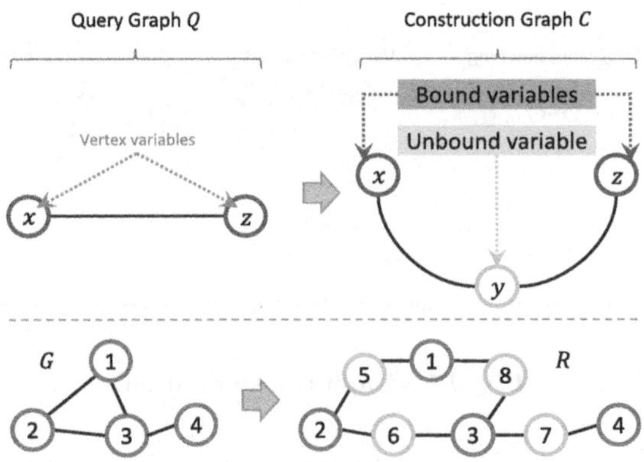

Fig. 18. Example of a graph transformation

where u_t^v is a new vertex uniquely created for t and v. The construction result is a graph $R(V_R, E_R)$, where $V_R = \{\pi(t, v) \mid t \in T \wedge v \in V_C\}$ and $E_R = \{(\pi(t, u), \pi(t, v)) \mid t \in T \wedge (u, v) \in E_C\}$.

Figure 18 shows a simple graph transformation. The query graph matches pairs of vertices x and z connected with an edge. The construction creates for any of these pairs a new vertex y and connects it to x and z with an edge, respectively. Note that the transformation does not result in a set of construction graph instantiations but a single (potentially partitioned) graph R. To illustrate this the figure also shows a graph G and the graph R that results from applying the transformation to G. The numbers in the vertices represent the vertex identities, clearly illustrating that vertex y in the construction graph C cause the transformation to create the four new vertices 5, 6, 7, and 8—one distinct vertex per match of query graph Q on the data graph G.

The definition of graph construction can be easily extended towards labeled graphs and property graphs. Generally, bound vertices in the construction graph are merely placeholders for vertices found by the matching. Hence, bound vertices do not have any labels or properties. In contrast, unbound vertices can have labels and properties. Property values in the construction graph are expressions from an expression system grounded on literals and bound variables, i.e. variable comprised in the schema of table T of variable bindings, cf. Definitions 16 and 17. More practically speaking, new property values are calculated from the literals and the properties of vertices bound in the current tuple t.

Figure 19 shows a graph transformation for a graph with labels. The transformation Γ is shown in pseudo code; the query graph Q, the construction graph C, an input graph G, and the resulting graph R are shown visually. Here, the query graph matches a person having two dogs d_1 and d_2. The construction adopts the two Dog vertices, neglects the dog owner and connects the two dogs with an edge labeled Buddies. Applied to graph G that features a person owning three dogs and one horse, results in a graph R that has the three dogs all connected as buddies.

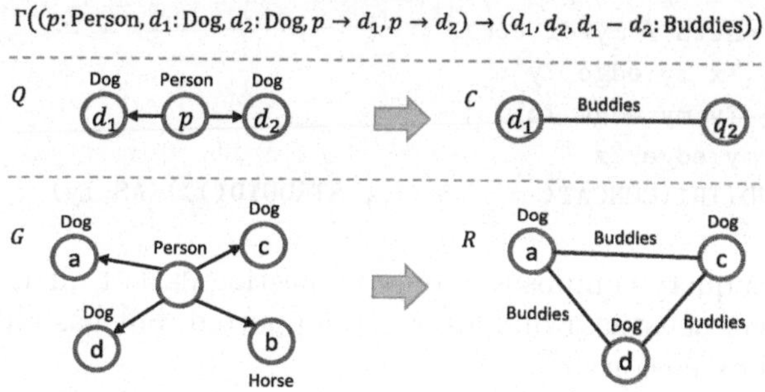

Fig. 19. Example of a graph transformation in a graph with labels.

The example illustrates that graph transformations do not necessarily have to produce new vertices, they can also be used to merely project the match result. Likewise, graph transformations do not necessarily have to adopt vertices from the input graph, they can be used to produce entirely new graphs. Further, graph transformations can be used to convert graphs between different meta models, i.e. between different graph types. In the example, the input graph is directed and has only vertex labels, while the resulting graph is undirected and features vertex and edge labels.

Among the query languages mentioned in Sect. 3.4, SPARQL and PGQL support the construction of graphs and, therefore, can express graph transformations. In SPARQL a query can have a `CONSTRUCT` clause in place of the `SELECT` clause. Similar to `WHERE`, `CONSTRUCT` takes a basic graph pattern (called template) which constitutes the construction graph for the graph construction. It is also possible to construct new vertices—either as blank node (vertices without IRI, an RDF specialty) or with the help of the `IRI` function. The transformation in Fig. 18 can be expressed using blank nodes as follows:

```
PREFIX my: <http://myrdf.com/>
CONSTRUCT { ?x  my:edge _:y .
            _:y my:edge ?z .  }
WHERE { ?x my:edge ?z . }
```

Using the `IRI` function is more complicated, because it has to construct a distinct IRI per match. One way to do that is to construct a new IRI from a UUID [38] string. UUID is a standardized procedure to create unique identifiers in distributed system without centralized coordination. A generated UUID is a 128-bit number typically represented by 32 lowercase hexadecimal digits. SPARQL provides a built-in function `STRUUID` to generate an UUID string. With the built-in functions `CONCAT` and `IRI`, the generated UUID string can be completed to an IRI string and type casted to an proper IRI, respectively. The `BIND` clause allows to assign the constructed IRI to a variable in the match pattern, so that this variable will be bound to a distinct IRI per match.

```
PREFIX my: <http://myrdf.com/>
CONSTRUCT { ?x my:edge ?y .
            ?y my:edge ?z . }
WHERE { ?x my:edge ?z
        BIND(IRI(CONCAT('my:newY/', STRUUID())) AS ?y) . }
```

SPARQL allows query composition only by nesting **SELECT** queries in the **WHERE**. Composition over **CONSTRUCT** queries is not supported, but has already been proposed as useful extension [39].

In PGQL, the **SELECT** clause offers special functions allowing graph construction. The constructor function **GRAPH** creates one graph per match based on a construction pattern. The aggregation function **UNION** merges multiple graphs into a single one. In PGQL, the example query looks like this:

```
SELECT UNION(GRAPH { x-[]-z,
                     z-[]-y }) AS newGraph
FROM someGraph
WHERE (x)-[]-(y)
```

New vertices can also be assigned with properties. PGQL allows also query composition over graphs by nesting queries that construct a graph in the **FROM** clause of the outer query.

A graph query language including graph transformations is also discussed and studied in [40]. Graph transformations have been proposed and discussed also in other contexts, e.g. software engineering [41–43].

4.2 Aggregation

As we have seen in the previous section, graph transformations are a powerful concept to change the structure of graphs and facilitate composability. However, graph construction happens per match. While it is possible to condense the matched subgraphs into smaller ones, it is not possible to have data condensation across multiple matches. This requires aggregation.

Fig. 20. Graph query with aggregation.

For a very simple example of an aggregation, consider a query that counts the number of edges of each vertex—the vertex's degree. Figure 20 illustrates such a query. For each vertex in the data graph the query yields a vertex with a new property showing the degree of the corresponding vertex in the data graph. Such a query requires aggregation—counting the neighboring vertices—and cannot be expressed with a standard non-aggregative graph transformation.

Aggregative graph transformations are similar to non-aggregative graph transformations, with the exception that the graph construction is based on a grouping construction graph. In the grouping construction graph, vertices have grouping variables assigned.

Definition 18 (Grouping Construction Graph). *A construction graph $C(V_C, E_C, \sigma_C, \gamma_C)$ for a table T of variable bindings is a graph where $\sigma_C : V_C \rightarrow \mathcal{X}$ is the injective variable assignment function and $\gamma_C : V_C \rightarrow \mathcal{P}(S(T))$ is the grouping function assigning sets of grouping variables to vertices in C. For every bound vertex $v \in V_C$ with $\sigma_C(v) \in S(T)$ constitutes its own grouping, so that $\gamma_C(v) = \sigma_C(v)$.*

Note that the grouping construction graph is a generalization of the non-grouping construction graph. With $\gamma(v) = S(T)$ for all unbound vertices v a grouping construction graph specifies exactly the same construction as its non-grouping counterpart. In any case grouping variables have to be bound in T.

Definition 19 (Aggregative Graph Construction). *Given are a table T of variable bindings in a graph $G(V, E)$ and a grouping construction graph $C(V_C, E_C, \sigma_C, \gamma_C)$. Let $\pi : T \times V_C \rightarrow \mathcal{V}$ be a function that maps a combination of binding tuple and construction graph vertex to a vertex out of all possible vertices $\mathcal{V}$, so that*

$$\pi(t, v) = \begin{cases} t[\sigma_C(v)], & if\ \sigma_C(v) \in S(T) \\ u^v_{t[\gamma_C(v)]}, & if\ \sigma_C(v) \notin S(T) \end{cases}$$

where $u^v_{t[\gamma_C(v)]}$ is a new vertex uniquely created for $t[\gamma_C(v)]$ and v and $t[\gamma_C(v)] = \{t[x] \mid x \in \gamma_C(v)\}$. The construction result is a graph $R(V_R, E_R)$, where $V_R = \{\pi(t, v) \mid t \in T \wedge v \in V_C\}$ and $E_R = \{(\pi(t, u), \pi(t, v)) \mid t \in T \wedge (u, v) \in E_C\}$.

The main difference of the aggregative graph construction to the non-aggregative one is how new vertices for unbound variables are created. Instead of creating them uniquely per tuple in T, they are created uniquely per grouping value set $t[\gamma_C(v)]$. The grouping value set of a construction vertex v is the projection of a tuple to v's grouping variables. For a group of tuples with identical bindings on grouping variables of a construction vertex, only one vertex for the whole group is constructed. Hence, aggregative graph construction allows to condense data across multiple matches.

With aggregative graph construction in the realm of property graphs, the expression system for calculating new property values can encompass aggregation functions such as COUNT, MIN, MAX, SUM, AVG, etc. to allow actual value aggregations. Note that by definition, edges in the construction graph implicitly inherit the grouping variables of the vertices they are attached to. In multigraphs edges may specify grouping variables additionally to the inherited grouping variables. In that case, for every edge $e \in E_C$ with $e = (u, v)$ it holds that $\gamma_C(u) \cup \gamma_C(v) \subseteq \gamma_C(e)$.

Let's reconsider the degree query sketched in Fig. 20. The query graph for this query is straight forward. It matches a vertex x with one of its neighbors y, $((x) - (y))$ in pseudo-code notation. For the grouping construction graph assume we list the grouping variables in pseudo-code notation behind the vertex variable separated by an @. For instance, $(a@b, c)$ denotes the construction vertex a with its grouping variable b and c. Further let's denote property expressions by listing <key>=<expression> pairs for the corresponding vertex. Using this notation, the grouping construction graph for the degree query is $(z@x\,(\mathsf{degree} = \mathtt{COUNT}(y)))$ and whole query is:

$$\Gamma\,(((x) - (y)) \rightarrow (z@x\,(\mathsf{degree} = \mathtt{COUNT}(y))))$$

The query matches all pairs of vertices x and y connected by an edge. For each group of such x–y pairs with the same vertex x the query creates a new vertex z and calculates the property **degree** for z by counting all vertices y in the corresponding group of matched x–y pairs. Because the pairs are grouped by x, the query counts exactly the number of neighbors of x, yielding the result sketched in Fig. 20. Since aggregative graph transformations are composable like non-aggregative graph transformations, we could run another query on that result to determine the average vertex degree in the original graph, for instance.

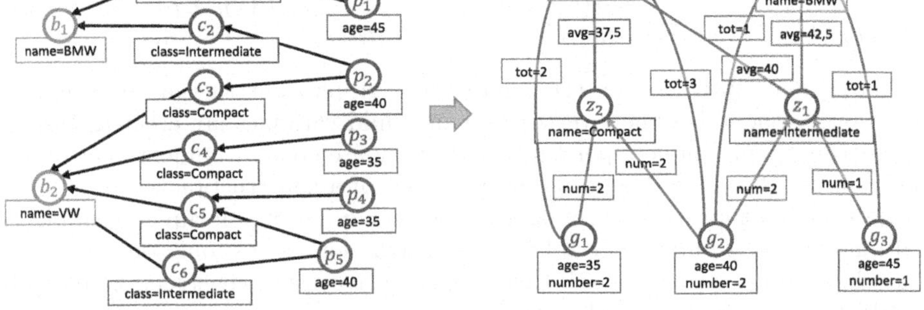

Fig. 21. Example of an aggregative graph transformation in a properties graph.

Figure 21 shows a more complex graph transformation with aggregation on a property graph. The transformation Γ is shown in pseudo code; an input graph G and the resulting graph R are shown visually; labels are represented by color. The query part of the transformation matches a chain of three vertices: a person, a car, and a car maker brand. Note that matching variables a and k are bound to property values. The construction keeps the matched Brand vertices and surrounds them with vertices representing different groups. For instance, AgeGroup vertices represent groups of equally old persons, e.g. g_2 is constructed for the group of matches where $a = 40$. With the aggregation function COUNT the number of matches in an a-group is assigned to the number property of the corresponding age group vertex—for instance number $= 2$ for g_2. The edges constructed between the new vertices inherit the vertices' grouping. For instance, edges $g \rightarrow z$ inherit a grouping on a and k. The edge (g_2, z_2) represents the group of matches where $a = 40$ and $k =$ Compact. Whereas the edge (z_2, b_2) represents the group of matches where $k =$ Compact and $b = b_2$. Remember that by definition bound vertices are grouped by themselves, so that edge $z \rightarrow b$ inherit b as a grouping variable. There are three matches in the Compact–b_2 group and the corresponding persons have an average age of 37.5, which is assigned as a property to that edge.

Aggregative graph transformations cannot be expressed in SPARQL. Although SPARQL has aggregation functions, they are not allowed in the `CONSTRUCT` clause. For PGQL no clear statement in that regard can be made as of now. The use of aggregation functions within the `GRAPH` function are undescribed as well as not explicitly ruled out.

5 Multidimensional Graph Queries

An important part of a data analyst's toolbox is multidimensional analytics. It is well understood and results can be easily communicated to business colleagues who are familiar with multidimensional reporting. Core element of the multidimensional analytics are multidimensional cubes (or just cubes for short). A cube provides a structured way to analyze a set of facts. It organizes the facts in dimensions, typically answers to questions about the facts of the sort: Who, what, where, when, why, and how. For instance, the question in which store a sale took place constitutes the dimension location for sales facts. A dimension can be hierarchical, i.e. there exists a hierarchy that categorizes the dimension's elements (the possible answers to the question presented by the dimension). Next to facts and dimensions, measures are associated with a cube. A measure is typically numerical information of the facts that can be aggregated along the dimensions of the cube. A multidimensional query calculates the measures of a cube at a specific setting of the cube's dimensions. For instance, the price of all sales can be summed up by store and month to see how much each store contributes to the total revenue.

With the increasing abundance of graph data, it becomes important to facilitate multidimensional analytics also on graphs. Graphs are not in the multidimensional model, but the data they contain can be analyzed through multidimensional glasses. In other words, we use a multidimensional cube as a prisma to look at the data graph. Naturally, what we want to see is a graph, more precisely the data graph aggregated according to the prisma we are looking through. One glimpse through the prisma with a specific dimensional setting is a multidimensional graph query. Since multidimensional graph queries produce graphs, they can be also thought of as graph transformations and are also composable.

We can distinguish two kinds of multidimensional analysis on graphs. The first kind is interested in entities represented in a graph; let's call this *entity-oriented multidimensional analysis*. Here the facts are individual vertices or subgraphs; what the facts exactly are depends on the intent of the data analyst, naturally. The second kind is interested in the structure; let's call this *structure-oriented multidimensional analysis*. Here the adjacency information of the graph constitutes the facts. In the following, we consider multidimensional queries of both kinds.

5.1 Entity-Oriented Multidimensional Queries

In the entity-oriented multidimensional analysis vertices or more generally subgraphs constitute the facts of interest. Of what we have learned in Sect. 3, subgraphs are declaratively queried with the help of patterns. Hence, the first ingredient of an entity-oriented multidimensional query (EMQ) is a fact pattern.

Definition 20 (Fact Pattern). *A fact pattern $F(V_F, E_F)$ on a data graph $G(V_G, E_G)$ is a graph that can be matched against G.*

In the data graph, a fact is connected by edges or paths to its dimensional information. For instance, the Sale vertex has an edge pointing to the sales location, another pointing to the product, etc. For simplicity and without loss of generality, we assume the fact pattern to encompass all the dimensional information of interest around the fact. Hence, a dimension is defined as a subset of the vertices in the fact pattern.

Definition 21 (Dimension in EMQs). *A dimension $d \subseteq V_F$ of a fact pattern $F(V_F, E_F)$ for a data graph $G(V_G, E_G)$ is a subset of vertices from the fact pattern F. Let $\sigma(d) = \{\sigma(v) \mid v \in d\}$ be the dimension instance of a specific match σ of F on G, i.e. the set of vertices in the match corresponding to the dimension vertices d in F according to σ. Let $\lambda(\sigma(d))$ be the label of the dimension instance $\sigma(d)$.*

Naturally, a measure is a calculation procedure that extracts the desired information from the facts and aggregates it.

Definition 22 (Measure in EMQs). *A measure $m(\sigma(V_F)) = a$ is an aggregation function calculating an aggregate a on a subgraph match σ. Let $\lambda(m(\sigma))$ be the label of the aggregate calculated by m on σ.*

Subsequently, an EMQ is defined by a fact pattern, dimensions, and a measure.

Definition 23 (Entity-oriented Multidimensional Query). *An entity-oriented multidimensional query Q is a triple (F, D, m) where F is a fact pattern, $D = \{d_1, \ldots, d_n\}$ is a set of dimensions, and m is a measure.*

For clarity, we limit the query to a single measure. The definition can be easily generalized to allow multiple measures. For the same reason we assume that the result of an EMQ is a labeled graph. Labels are not a necessity, properties could be used likewise.

The result of an EMQ is the result of an aggregative graph transformation. The fact pattern is the query graph of the transformation, while the dimensions and the measure define the grouping construction graph. Essentially, the construction graph contains a vertex for every dimension and one vertex for the measure.

Definition 24 (Entity-oriented Multidimensional Query Semantics). *Given a data graph $G(V_G, E_G)$ and an entity-oriented multidimensional query $Q(F, D, m)$, let $\sigma : V_F \rightarrow V_G$ be a subgraph match of F on G. A labeled graph $R(V_R, E_R, \Sigma_R, \lambda_R)$ is the result for Q on G such that R is the result of an aggregative graph transformation $\Gamma_{F,C}$ where the $C(V_C, E_C, \lambda_C, \sigma_C, \gamma_C)$ is a labeled grouping construction graph such that*

- *there is a vertex for every dimension d labeled with $\lambda(\sigma(d))$ and a vertex for the measure m labeled with the aggregate, i.e. $V_C = \{v_d \mid d \in D\} \cup \{v^m\}$ and $\lambda_C(v_d) = \lambda(\sigma(d))$ and $\lambda_C(v_m) = \lambda(m(\sigma))$;*
- *there is an unlabeled edge from the measure vertex v_m to every dimension vertex v_d, i.e. $V_E = \{(v_m, v_d) \mid d \in D\}$;*
- *dimension vertices are grouped according to the dimension's definition, so that $\gamma_C(v_d) = d$ for every $d \in D$, and the measure vertex is group according to the definitions of all dimensions, so that $\gamma_C(v_m) = \bigcup_D d$; and*
- *all dimension vertices and the measure vertex are unbound and have an arbitrary but unique variable assigned.*

Note that the labels in the construction graph depend on σ. During graph construction, when the construction graph gets instantiated for every match σ, the labels are instantiated for each σ. Dimension vertices representing single-vertex dimensions with $|d| = 1$ could also be realized as bound vertices, such that $\{v_d\} = d$ and $\gamma_C(v_d) = \sigma_C(v_d)$.

Figure 22 shows an example of an EMQ on a property graph. The query $Q(F, \{b, k, a, l\}, \texttt{COUNT}(c))$ counts cars c (the facts) in groups of brand b, vehicle class k, owner age a, and registration country l (the dimensions). As can be seen, the fact pattern queries the subgraph around a Car vertex that encompasses all the desired dimension information. The dimension information can be a property of the fact vertex, such as class, an adjacent vertex, such as a Brand vertex, or indirectly connected elements, such as the age property of the owner or the Country vertex that is two hops away. The figure also shows a data graph G, the aggregative graph transformation Γ that implements the query, and the result graph R of the query on G.

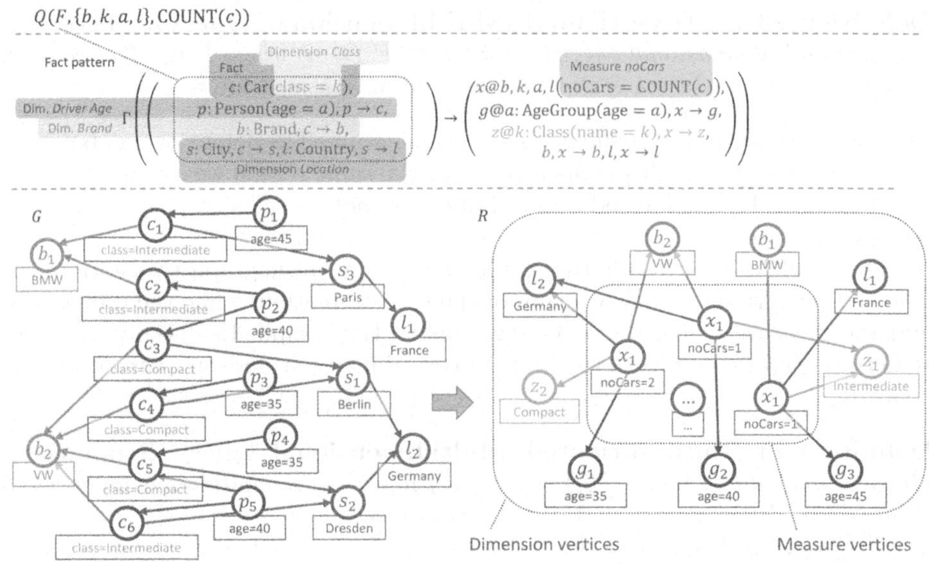

Fig. 22. Example of entity-oriented multidimensional query on a properties graph.

The principles of EMQs have been employed in [44–46] for declarative interactive graph analytics. Here, the cube definition exists outside the graph and references the graph with the help of patterns. Another approach, particularly for RDF data, is to annotate the data in the graph so that the cube definition becomes part of the graph. A number of RDF vocabularies has been proposed for that purpose, e.g. [47–49]. For an extensive discussion for OLAP on RDF data please see [50].

5.2 Structure-Oriented Multidimensional Queries

In the structure-oriented multidimensional graph analysis the adjacency constitutes the facts of interest. Accordingly, the facts are not specified by the query but are implicitly given. Dimensions are vertex properties and edge properties used to group vertices and edges, respectively. Likewise labels can also be used as dimensions. Basically, a structure-oriented multidimensional query (SMQ) summarizes a graph's structure according to different dimensions provided by the user.

For a simple example consider Fig. 23. The illustrated query summarizes the structure of the graph on the left into the graph on the right—often called a summary graph. The query uses the vertex color as dimension and a simple `COUNT` of vertices and edges as measure. For simplicity numerical quantities on vertices and edges are represented by vertex size and edge thickness, respectively. As can be seen, the resulting summary graph informs us intuitively that there are more red vertices than blue vertices in the original data graph. It also shows

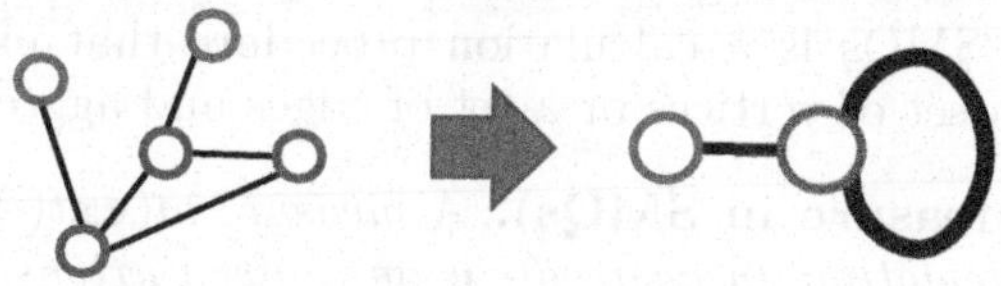

Fig. 23. Structure-oriented multidimensional query. (Color figure online)

that red vertices are heavily connected among themselves. In contrast, blue vertices are only connected to red vertices but have no edges among each other. Structure-oriented multidimensional queries are very useful when exploring or analyzing graphs. With vertex and edge labels used as dimensions, structure-oriented multidimensional queries provide a concise representation of the data graph's schema. This is particularly useful for visual graph exploration interface because the resulting summary graph quickly informs a user what data she is dealing with.

For the definitions in the following we assume a property graph $G(V, E, K, \mathcal{D}, \pi)$ without labels and multiedges (cf. Definition 3). Generally, SMQs distinguish between vertex dimensions and edge dimensions, although both are very similar.

Definition 25 (Vertex Dimension in SMQs). *A vertex dimension $d^V \subset K_G$ in $G(V_G, E_G, K_G, \mathcal{D}_G, \pi_G)$ is a set of property keys. Additionally:*

- *Let $\pi_G(v, d^V) = \{\pi_G(v, k) \mid k \in d^V\}$ be the vertex dimension instance of a vertex $v \in V_G$, i.e. the property values of v on dimension d^V.*
- *Let $\mathit{img}(d^V) = \{\pi_G(v, d^V) \mid v \in V_G\}$ be the set of all vertex dimension instances of dimension d^V in G.*
- *Let $d^V_x \in \mathit{img}(d^V)$ be a vertex dimension instance of dimension d^V and $\lambda(d^V_x)$ be the label of that instance.*
- *Let $\mathit{ext}(d^V_x) = \{v \in V_G \mid \pi_G(v, d^V) = d^V_x\}$ be the extension of vertex dimension instance d^V_x, i.e. all vertices whose dimension instance is d^V_x.*

Definition 26 (Edge Dimension in SMQs). *An edge dimension $d^E \subset K_G$ in $G(V_G, E_G, K_G, \mathcal{D}_G, \pi_G)$ is a set of property keys. Additionally:*

- *Let $\pi_G(e, d^E) = \{\pi_G(e, k) \mid k \in d^E\}$ be the edge dimension instance of an edge $e \in E_G$, i.e. the property values of e on dimension d^E.*
- *Let $\mathit{img}(d^E) = \{\pi_G(e, d^E) \mid e \in E_G\}$ be the set of all edge dimension instances of dimension d^E in G.*
- *Let $d^E_x \in \mathit{img}(d^E)$ be an edge dimension instance of dimension d^E and $\lambda(d^E_x)$ be the label of that instance.*
- *Let $\mathit{ext}(d^V_x, d^E_y, d^V_z) = \{e \in E_G \mid \pi_G(\mathit{src}(e), d^V) = d^V_x \land \pi_G(e, d^E) = d^E_y \land \pi_G(\mathit{trg}(e), d^V) = d^V_z\}$ be the extension of a combination of vertex dimension instances d^V_x and d^V_z and edge dimension instance d^E_y, i.e. the set of all edges e with dimension instance d^E_y and a source and a target vertex with dimension instance d^V_x and d^V_z, respectively.*

A measure for SMQs is a calculation procedure that extracts the desired information from a set of vertices or a set of edges and aggregates it.

Definition 27 (Measure in SMQs). *A measure* $m(\mathtt{ext}(\cdot)) = a$ *is an aggregation function calculating an aggregate* a *on a given extension, i.e. on a set of vertices or on a set of edges. Let* $\lambda(m(\mathtt{ext}(\cdot)))$ *be the label of the aggregate.*

An SMQ is defined by a vertex dimension, an edge dimension, a vertex measure, and an edge measure.

Definition 28 (Structure-oriented Multidimensional Query). *A structure-oriented multidimensional query* Q *is a tuple* (d^V, d^E, m^V, m^E) *where* d^V *is a vertex dimension,* d^E *is an edge dimension,* m^V *is a vertex measure, and* m^E *is an edge measure.*

For clarity, we limit the query to a single dimension and a single measure for vertices and edges, respectively. The definition can be easily generalized to allow multiple dimensions and multiple measures. For the same reason, we assume the results of an SMQ to be a labeled multigraph. Labels are not a necessity, properties could be used likewise.

The result of an SMQ is a graph with a representative vertex for each vertex group in the data graph and an representative edge for each edge group in the data graph. Edges are grouped by the dimension of their source vertex, the dimension of their target vertex, and the edge dimension. This way, the structure of the result graph is a summary of the data graph.

Definition 29 (Structure-oriented Multidimensional Query Semantics). *Given a data graph* $G(V_G, E_G, K_G, \mathcal{D}_G, \pi_G)$ *and a structure-oriented multidimensional query* $Q(d^V, d^E, m^V, m^E)$*, a labeled multigraph* $R(V_R, E_R, \rho_R, \Sigma_R, \lambda_R)$ *is the result for* Q *on* G *such that*

- *there is a vertex* v_x *for every dimension instance* $d^V_x \in \mathtt{img}(d^V)$ *of the vertex dimension* d_V *labeled with* $\lambda(d^V_x) \cdot \lambda(m^V(\mathtt{ext}(d^V_x)))$*; and*
- *there is an edge* $e_{x,y,z}$ *for every non-empty extension* $X = \mathtt{ext}(d^V_x, d^E_y, d^V_z)$ *with* $|X| \neq \emptyset$ *labeled with* $\lambda(d^E_y) \cdot \lambda(m^E(X))$ *and incident to* $\rho_R(e_{x,y,z}) = (v_x, v_z)$.

For both, vertices and edges in the result graph, the labels are a concatenation of the dimension instance label and the aggregate label. This is merely a simplification for the sake of clarity. In a property graph, dimension instance and aggregate could be easily attached as separated properties to vertices and edges.

SMQ semantics can be defined also with the help of graph transformations. An SMQ can be represented with two graph transformations and a union. The first transformation

$$\Gamma_V\left((v(d^V = x)) \rightarrow (v_x @ x : \mathsf{Group}(\mathsf{dim} = x, \mathsf{agg} = m^V(x)), v, v_x \rightarrow v)\right)$$

aggregates vertices bound to variable v according to their vertex dimension instance x to group vertices v_x. Here, dimension instance and aggregate are attached as properties to the new vertex. The transformation also links $(v_x \rightarrow v)$ group vertices x_g to their group members x, i.e. to all vertices in $\mathtt{ext}(g)$. The result of transformation Γ_V is united with the original graph and the second transformation is applied to that union to produce the query result. The second transformation

$$\Gamma_E\left(\left(\begin{array}{c} v_x:\mathsf{Group}, v_x\rightarrow v,\\ u_z:\mathsf{Group}, u_z\rightarrow u,\\ v\rightarrow u(d_E=y)\end{array}\right)\rightarrow\left(\begin{array}{c} v_x, u_z,\\ v_x\rightarrow u_z@h(\mathsf{dim}=y,\mathsf{agg}=m^E(v\rightarrow u))\end{array}\right)\right)$$

aggregates the edges between original vertices v and u grouped by group vertices v_x and u_z into edges between the group vertices v_x and u_z. Again, dimension instance and aggregate are attached as properties.

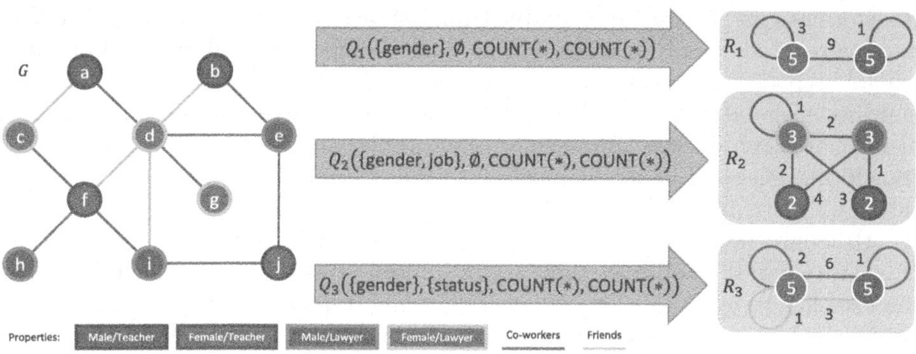

Fig. 24. Three examples of structure-oriented multidimensional queries.

Figure 24 shows the three different SMQs Q_1, Q_2, and Q_3 on data graph G and their respective result graphs R_1, R_2, and R_3. All three queries summarize G's structure, along different dimensions, though. The measure for vertices and for edges is a simple `COUNT` in all queries, so that the result graph informs how many vertices or edge are in the respective group. For instance, R_1 shows that their are 5 female vertices and 5 male vertices and 9 edges between a female–male vertex pair in G. Note, that edges in G between vertices that are in the same group, i.e. that have the same dimension instance, produce a loop edge in the result graph. As can be also seen in Q_1, a dimension may also be an empty set. An empty vertex dimension causes all vertices to fall in the same group; an empty edge dimension (as in Q_1) causes edges to be grouped only by the groups of their source and target vertices.

SMQs have be proposed as *cuboid queries* in [51], without grouping of edges, though. Assuming all combinations of property keys in a given graph make a reasonable dimension, [51] defines a *graph cube* as the set of all SMQs that can be formulated on a graph. The notion of graph cube is also considered in [52]. SMQs on property graphs, including grouping of edges, are consider in [53].

6 Conclusion

Today, graph data can be found in an enormous heterogeneity of domains and applications, combined with an increased desire and need to analyze it. Descriptive multidimensional data analytics is the most fundamental form of analytics and ought to receive out-of-the-box support by graph database management systems. One pivotal conceptual issue in this regard is the formulation of multidimensional graph queries in a declarative graph query language. In this article, we took an elaborated look formulating declarative multidimensional graph queries based on subgraph matching. The focus was on query language definition and semantics; query processing was not considered. Since it is the common foundation of most declarative graph query languages, we had a detailed discussion of subgraph matching. Building on that, we considered the extension of subgraph matching towards graph transformations and aggregative graph transformations. Aggregative graph transformations allow to condense graphs into aggregated, more abstract graphs while being composable over graphs—a crucial step towards declarative analytics. Finally, we considered two kinds of multidimensional graph queries: Entity-oriented and structure-oriented multidimensional graph queries. As we have seen, both can be defined on top of graph transformations and hence fit nicely into the system built on top of the idea of subgraph matching.

Although the discussed analytical queries are very useful and essential part of a data analyst's toolbox, they are of limited power. To get insightful results the user has to formulate precise queries, which typically requires already some deeper understanding of considered data. For instance, to formulate a structure-oriented multidimensional graph query the user has to choose dimensions and measures. In cases where user has no intuition about the data, it will be hard to formulate queries that produce new and interesting insights about the data. More advanced methods, which e.g. discover interesting dimension–measure settings themselves, are not yet investigated regarding their integration in declarative graph query languages. Extending graph query languages towards the domain of advanced graph analytics is a research challenge for the upcoming years.

References

1. Berners-Lee, T., Hendler, J., Lassila, O.: The semantic web. Sci. Am. **284**(5), 34–43 (2001). https://www.scientificamerican.com/magazine/sa/2001/05-01/
2. Auer, S., Bizer, C., Kobilarov, G., Lehmann, J., Cyganiak, R., Ives, Z.: DBpedia: a nucleus for a web of open data. In: Aberer, K., et al. (eds.) ASWC/ISWC - 2007. LNCS, vol. 4825, pp. 722–735. Springer, Heidelberg (2007). doi:10.1007/978-3-540-76298-0_52
3. Franklin, M.J., Halevy, A.Y., Maier, D.: From databases to dataspaces: a new abstraction for information management. SIGMOD Rec. **34**(4), 27–33 (2005)

4. Werner, H., Bornhövd, C., Kubis, R., Voigt, H.: MOAW: an agile visual modeling and exploration tool for irregularly structured data. In: Datenbanksysteme für Business, Technologie und Web (BTW), BTW 2011, 14. Fachtagung des GI-Fachbereichs Datenbanken und Informationssysteme (DBIS), Kaiserslautern, Germany, 2–4 March 2011. LNI, vol. 180, pp. 742–745. GI (2011)
5. Beck, K., Beedle, M., van Bennekum, A., Cockburn, A., Cunningham, W., Fowler, M., Grenning, J., Highsmith, J., Hunt, A., Jeffries, R., Kern, J., Marick, B., Martin, R.C., Mellor, S., Schwaber, K., Sutherland, J., Thomas, D.: Manifesto for Agile Software Development (2001). http://agilemanifesto.org/
6. Hey, T., Tansley, S., Tolle, K.M.: The Fourth Paradigm: Data-Intensive Scientific Discovery. Microsoft Research, Redmond (2009)
7. Chan, P., Shoshani, A.: SUBJECT: a directory driven system for organizing and accessing large statistical databases. In: Proceedings of the 7th International Conference on Very Large Data Bases, VLDB 1981, 9–11 September 1981, Cannes, France, pp. 553–563. IEEE (1981)
8. Angles, R., Gutiérrez, C.: Survey of graph database models. ACM Comput. Surv. **40**(1), 1:1–1:39 (2008)
9. Angles, R.: A comparison of current graph database models. In: 3rd International Workshop on Graph Data Management, GDM 2012, Washington DC, USA, 5 April 2012, pp. 171–177. IEEE (2012)
10. W3C: RDF 1.1 Concepts and Abstract Syntax, February 2014. http://www.w3.org/TR/2014/REC-rdf11-concepts-20140225/
11. Dürst, M., Suignard, M.: Internationalized Resource Identifiers (IRIs), RFC 3987, January 2005. http://tools.ietf.org/html/rfc3987
12. Berners-Lee, T., Fielding, R.G., Masinter, L.: Uniform Resource Identifier (URI): Generic Syntax, RFC 3986, January 2005. http://tools.ietf.org/html/rfc3986
13. Rodriguez, M.A., Neubauer, P.: Constructions from dots and lines. Bull. Am. Soc. Inf. Sci. Technol. **36**(6), 35–41 (2010)
14. van Rest, O., Hong, S., Kim, J., Meng, X., Chafi, H.: PGQL: a property graph query language. In: Proceedings of the Fourth International Workshop on Graph Data Management Experiences and Systems, GRADES 2016, Redwood Shores, CA, USA, 24–24 June 2016, p. 7. ACM (2016)
15. Bornhövd, C., Kubis, R., Lehner, W., Voigt, H., Werner, H.: Flexible information management, exploration and analysis in SAP HANA. In: Proceedings of the International Conference on Data Technologies and Applications, DATA 2012, Rome, Italy, 25–27 July 2012, pp. 15–28. SciTePress (2012)
16. Rudolf, M., Paradies, M., Bornhövd, C., Lehner, W.: The graph story of the SAP HANA database. In: Proceedings of the Datenbanksysteme für Business, Technologie und Web (BTW), BTW 2013, 15. Fachtagung des GI-Fachbereichs Datenbanken und Informationssysteme (DBIS), Magdeburg, Germany, 11–15 March 2013. LNI, vol. 214, pp. 403–420. GI (2013)
17. Martínez-Bazan, N., Muntés-Mulero, V., Gómez-Villamor, S., Nin, J., Sánchez-Martínez, M., Larriba-Pey, J.: DEX: high-performance exploration on large graphs for information retrieva. In: Proceedings of the Sixteenth ACM Conference on Information and Knowledge Management, CIKM 2007, Lisbon, Portugal, 6–10 November 2007, pp. 573–582. ACM (2007)
18. Martínez-Bazan, N., Gómez-Villamor, S., Escale-Claveras, F.: DEX: a high-performance graph database management system. In: 2nd International Workshop on Graph Data Management, GDM 2011, Hannover, Germany, 11–16 April 2011, pp. 124–127. IEEE (2011)

19. Ullmann, J.R.: An algorithm for subgraph isomorphism. J. ACM **23**(1), 31–42 (1976)
20. Henzinger, M.R., Henzinger, T.A., Kopke, P.W.: Computing simulations on finite and infinite graphs. In: 36th Annual Symposium on Foundations of Computer Science, Milwaukee, Wisconsin, 23–25 October 1995, pp. 453–462 (1995)
21. Miller, J.A., Ramaswamy, L., Kochut, K.J., Fard, A.: Research directions for big data graph analytics. In: 2015 IEEE International Congress on Big Data, BD 2015, New York City, NY, USA, 27 June–2 July 2015, pp. 785–794. IEEE (2015)
22. Fan, W., Li, J., Ma, S., Tang, N., Wu, Y., Wu, Y.: Graph pattern matching: from intractable to polynomial time. Proc. VLDB Endow. **3**(1), 264–275 (2010)
23. Wood, P.T.: Query languages for graph databases. SIGMOD Rec. **41**(1), 50–60 (2012)
24. Hellings, J., Kuijpers, B., den Bussche, J.V., Zhang, X.: Walk logic as a framework for path query languages on graph databases. In: Proceedings of the 16th International Conference on Database Theory, ICDT 2013, Genoa, Italy, 18–22 March 2013, pp. 117–128. ACM (2013)
25. Figueira, D., Libkin, L.: Path logics for querying graphs: combining expressiveness and efficiency. In: 30th Annual ACM/IEEE Symposium on Logic in Computer Science, LICS 2015, Kyoto, Japan, 6–10 July 2015, pp. 329–340. IEEE (2015)
26. Barcelo, P., Fontaine, G., Lin, A.W.: Expressive path queries on graph with data. Logical Methods Comput. Sci. **11**(4), 1–39 (2015)
27. Libkin, L., Martens, W., Vrgoc, D.: Querying graphs with data. J. ACM **63**(2), 14 (2016)
28. Barceló, P.: Querying graph databases. In: Proceedings of the 32nd ACM SIGMOD-SIGACT-SIGART Symposium on Principles of Database Systems, PODS 2013, New York, NY, USA, 22–27 June 2013, pp. 175–188. ACM (2013)
29. Bourhis, P., Krötzsch, M., Rudolph, S.: Query containment for highly expressive datalog fragments. The Computing Research Repository abs/1406.7801, June 2014
30. Reutter, J.L., Romero, M., Vardi, M.Y.: Regular queries on graph databases. In: 18th International Conference on Database Theory, ICDT 2015, Brussels, Belgium, 23–27 March 2015. Leibniz International Proceedings in Informatics (LIPIcs), vol. 31. Schloss Dagstuhl-Leibniz-Zentrum fuer Informatik, Dagstuhl, pp. 177–194 (2015)
31. W3C: SPARQL 1.1 Overview, March 2013. http://www.w3.org/TR/2013/REC-sparql11-overview-20130321/
32. W3C: SPARQL 1.1 Query Language, March 2013. https://www.w3.org/TR/2013/REC-sparql11-query-20130321/
33. Nenov, Y., Piro, R., Motik, B., Horrocks, I., Wu, Z., Banerjee, J.: RDFox: a highly-scalable RDF store. In: Arenas, M., et al. (eds.) ISWC 2015. LNCS, vol. 9367, pp. 3–20. Springer, Cham (2015). doi:10.1007/978-3-319-25010-6_1
34. Arenas, M., Conca, S., Pérez, J.: Counting beyond a Yottabyte, or how SPARQL 1.1 property paths will prevent adoption of the standard. In: Proceedings of the 21st World Wide Web Conference 2012, WWW 2012, Lyon, France, 16–20 April 2012, pp. 629–638. ACM (2012)
35. Rodriguez, M.A.: The Gremlin graph traversal machine and language. In: ACM Database Programming Languages Conference, DBPL 2015. ACM (2015)
36. Date, C.J.: Some principles of good language design (with especial reference to the design of database languages). SIGMOD Rec. **14**(3), 1–7 (1984)
37. Abrahamsen, M., Eilers, S.: On the asymptotic enumeration of LEGO structures. Exp. Math. **20**(2), 145–152 (2011)

38. Leach, P., Mealling, M., Salz, R.: A Universally Unique IDentifier (UUID) URN Namespace, RFC 4122, July 2005. https://tools.ietf.org/html/rfc4122
39. Angles, R., Gutierrez, C.: Subqueries in SPARQL. In: Proceedings of the 5th Alberto Mendelzon International Workshop on Foundations of Data Management, AMW 2011, Santiago, Chile, 9–12 May 2011, vol. 749. CEUR-WS.org (2011)
40. Martín, M.S., Gutierrez, C., Wood, P.T.: SNQL: a social networks query and transformation language. In: Proceedings of the 5th Alberto Mendelzon International Workshop on Foundations of Data Management, AMW 2011, Santiago, Chile, 9–12 May 2011 (2011)
41. Schürr, A.: Specification of graph translators with triple graph grammars. In: Mayr, E.W., Schmidt, G., Tinhofer, G. (eds.) WG 1994. LNCS, vol. 903, pp. 151–163. Springer, Heidelberg (1995). doi:10.1007/3-540-59071-4_45
42. Königs, A., Schürr, A.: Tool integration with triple graph grammars - a survey. Electron. Notes Theor. Comput. Sci. **148**(1), 113–150 (2006)
43. Heckel, R.: Graph transformation in a nutshell. Electron. Notes Theor. Comput. Sci. **148**(1), 187–198 (2006)
44. Rudolf, M., Voigt, H., Bornhövd, C., Lehner, W.: SynopSys: foundations for multidimensional graph analytics. In: Castellanos, M., Dayal, U., Pedersen, T.B., Tatbul, N. (eds.) BIRTE 2013-2014. LNBIP, vol. 206, pp. 159–166. Springer, Heidelberg (2015). doi:10.1007/978-3-662-46839-5_11
45. Rudolf, M., Voigt, H., Bornhövd, C., Lehner, W.: Multidimensional graph analytics. United States Patent, 20,160,063,081, 3 March 2016
46. Rudolf, M., Voigt, H., Lehner, W.: SPARQLytics: multidimensional analytics for RDF. In: Proceedings of the Datenbanksysteme für Business, Technologie und Web (BTW), BTW 2017, 15. Fachtagung des GI-Fachbereichs Datenbanken und Informationssysteme (DBIS), Stuttgart, Germany, 6–10 March 2017. GI (2017)
47. Etcheverry, L., Vaisman, A.A.: QB4OLAP: a vocabulary for OLAP cubes on the semantic web. In: Proceedings of the Third International Workshop on Consuming Linked Data, COLD 2012, Boston, MA, USA, 12 November 2012, vol. 905 (2012)
48. Matei, A., Chao, K.-M., Godwin, N.: OLAP for multidimensional semantic web databases. In: Castellanos, M., Dayal, U., Pedersen, T.B., Tatbul, N. (eds.) BIRTE 2013-2014. LNBIP, vol. 206, pp. 81–96. Springer, Heidelberg (2015). doi:10.1007/978-3-662-46839-5_6
49. W3C: The RDF Data Cube Vocabulary, January 2014. https://www.w3.org/TR/2014/REC-vocab-data-cube-20140116/
50. Abelló, A., Romero, O., Pedersen, T.B., Llavori, R.B., Nebot, V., Cabo, M.J.A., Simitsis, A.: Using semantic web technologies for exploratory OLAP: a survey. IEEE Trans. Knowl. Data Eng. **27**(2), 571–588 (2015)
51. Zhao, P., Li, X., Xin, D., Han, J.: Graph cube: on warehousing and OLAP multidimensional networks. In: Proceedings of the ACM SIGMOD International Conference on Management of Data, SIGMOD 2011, Athens, Greece, 12–16 June 2011, pp. 853–864. ACM (2011)
52. Wang, Z., Fan, Q., Wang, H., Tan, K., Agrawal, D., El Abbadi, A.: Pagrol: PArallel GRaph OLap over large-scale attributed graphs. In: Proceedings of the 30th International Conference on Data Engineering, ICDE 2014, Chicago, IL, USA, 31 March–4 April 2014, pp. 496–507. IEEE (2014)
53. Junghanns, M., Petermann, A., Rahm, E.: Distributed grouping of property graphs with GRADOOP. In: Proceedings of the Datenbanksysteme für Business, Technologie und Web (BTW), BTW 2017, 17. Fachtagung des GI-Fachbereichs Datenbanken und Informationssysteme (DBIS), Stuttgart, Germany, 6–10 March 2017 (2017)

Computational Approaches to Translation Studies

Shuly Wintner(✉)

Department of Computer Science, University of Haifa, Haifa, Israel
shuly@cs.haifa.ac.il

Abstract. Translated texts, in any language, have unique characteristics that set them apart from texts originally written in the same language. *Translation studies* is a research field that focuses on investigating these characteristics. Until recently, research in computational linguistics, and specifically in machine translation, has been entirely divorced from translation studies. The main goal of this tutorial is to introduce some of the findings of translation studies to researchers interested mainly in machine translation, and to demonstrate that awareness of these findings can result in better, more accurate machine translation systems (This chapter synthesizes material that has been previously published by the author and colleagues, in particular in Volansky et al. (2015); Rabinovich and Wintner (2015); Lembersky et al. (2011, 2012a, 2012b, 2013); and Twitto et al. (2015)).

1 Introduction

Research in *translation studies* reveals that translated texts are ontologically different from original, non-translated ones.[1] Translated texts, in any language, can be considered a "dialect" of that language, known as *translationese*. Several characteristics of translationese have been proposed as universal in a series of hypotheses. Awareness of the special properties of translationese can improve the quality of natural language processing (NLP) applications, in particular machine translation (MT). This chapter provides an introduction to translation studies and its relevance to research in NLP and specifically to MT.

In Sect. 2 we survey some theoretical hypotheses of translation studies. Focusing on the unique properties of translationese, we distinguish between properties resulting from *interference* of the source language (the so-called "fingerprints" of the source language on the translation product) and properties that are source-language-independent, and that are therefore presumably universal. The latter include phenomena resulting from three main processes: simplification, standardization and explicitation. All these phenomena are defined, explained and exemplified.

Section 3 describes several works that use standard (supervised and unsupervised) text classification techniques to distinguish between translations and

[1] We use 'originals' here as opposed to 'translations', although translation are of course also originally created by translators.

P. Marcel and E. Zimányi (Eds.): eBISS 2016, LNBIP 280, pp. 38–58, 2017.
DOI: 10.1007/978-3-319-61164-8_2

originals, in several languages. We focus on the features that best distinguish between the two classes, and how these features corroborate some (but not all) of the hypotheses set forth by translation studies scholars.

Then, we discuss in Sect. 4 several computational works that show that awareness of translationese can improve machine translation. Specifically, we show that language models compiled from translated texts are more fitting to the reference sets than language models compiled from originals. We also show that translation models compiled from texts that were (manually) translated from the source to the target are much better than translation models compiled from texts that were translated in the reverse direction. Finally, in Sect. 5 we touch upon some related issues and current research directions.

2 Translationese

Numerous studies suggest that translated texts differ from original ones. Gellerstam (1986) compared texts written originally in Swedish with texts translated from English into Swedish. He noted that the differences between them did not indicate poor translation but rather a statistical phenomenon, which he termed *translationese*.

The features of translationese were theoretically organized under the terms *laws of translation* or *translation universals*. Toury (1980, 1995) distinguished between two laws: the *law of interference* and the *law of growing standardization*. The former pertains to the fingerprints of the source text that are left in the translation product. The latter pertains to the effort to standardize the translation product according to existing norms in the target language and culture. The combined effect of these laws creates a hybrid text that partly corresponds to the source text and partly to texts written originally in the target language, but in fact is neither of them (Frawley 1984).

Baker (1993) suggested several candidates for translation universals, which she claimed to appear in any translated text, regardless of the source language: "features which typically occur in translated text rather than original utterances and which are not the result of interference from specific linguistic systems" (Baker 1993, p. 243). Among the better known universals are *simplification* and *explicitation*, defined and discussed thoroughly by Blum-Kulka and Levenston (1978, 1983) and Blum-Kulka (1986), respectively. A third universal hypotheses is *standardization*, also known as *normalization* (Toury 1995). We now detail these hypotheses.

Simplification refers to the process of rendering complex linguistic features in the source text into simpler features in the target text. Strictly speaking, this phenomenon can be studied only vis-à-vis the source text, since 'simpler' is defined here in reference to the source text, where, for example, the practice of splitting sentences or refraining from complex subordinations can be observed. And indeed, this is how simplification was first defined and studied in translation studies (Blum-Kulka and Levenston 1983; Vanderauwerea 1985). Baker (1993) suggested that simplification can be studied by comparing translated texts with non-translated ones, as long as both texts share the same domain, genre, time frame, etc. In a series of corpus-based studies, Laviosa (1998, 2002) confirmed this hypothesis. Ilisei et al. (2010) and Ilisei and Inkpen (2011) provided further evidence for this universal in Romanian and Spanish.

Explicitation is the tendency to spell out in the target text utterances that are more implicit in the source. One obvious way in which explicitation is manifested is by overusing cohesive markers such as *because*, *hence*, or *moreover*. Blum-Kulka (1986) exemplified this phenomenon in translations from Hebrew to English, and Øverås (1998) compiled a parallel bidirectional Norwegian-English and English-Norwegian corpus and provided further evidence for explicitation. Koppel and Ordan (2011) found that some of the prominent features in their list of function words were cohesive markers, such as *therefore*, *thus*, and *consequently*.

Translators take great efforts to *standardize* texts (Toury 1995), or, in the words of Baker (1993, p. 244), they have "a strong preference for conventional 'grammaticality'". This includes the tendency to avoid repetitions (Ben-Ari 1998), the tendency to use a more formal style manifested in refraining from the use of contractions (Olohan 2003), and the tendency to overuse fixed expressions even when the source text refrains, sometime deliberately, from doing so (Toury 1980; Kenny 2001).

In the last two decades corpora have been used extensively to study translationese. For example, Al-Shabab (1996) showed that translated texts exhibit lower lexical variety than originals; Laviosa (1998) showed that their mean sentence length is lower, as is their lexical density (ratio of content to non-content words). Both these studies provide evidence for the simplification hypothesis. Corpus-based translation studies became a very prolific area of research (Laviosa 2002).

3 Identification of Translationese

3.1 Supervised Classification

Since the pioneering work of Baroni and Bernardini (2006), text classification methods, based on standard machine learning techniques, have been extensively used to automatically identify translationese in several languages (van Halteren 2008; Ilisei et al. 2010; Ilisei and Inkpen 2011; Popescu 2011; Koppel and Ordan 2011; Avner et al. 2016). While many of these works were mainly interested in the practical task of distinguishing between originals and translations, Volansky et al. (2015) used the accuracy of classification as a proxy for evaluating the validity of translation studies hypotheses.

In *supervised machine-learning*, a *classifier* is trained on labeled examples the classification of which is known a priori, e.g., *translations* vs. *originals*. Each text has to be *represented*: a set of numeric *features* is extracted from the data (here, chunks of text), and a generic machine-learning algorithm is then trained to distinguish between *feature vectors* representative of one class and those representative of the other. Given enough data for training and given that the features are indeed relevant, the trained classifier can then be given an 'unseen' text, namely a text that is not included in the training set. Such a text is again represented by a feature vector in the same manner, and the classifier can predict the class (variety) it belongs to.

For evaluation, it is customary to use k-fold cross validation. The 'unseen' texts are also labeled, of course, and the prediction of the classifier can be compared to the actual, 'gold' label. In k-fold cross-validation, the training data is divided to k *folds* (typically, $k = 10$), and the following procedure is repeated k times: training on $k - 1$ folds, then testing on the held-out fold, cyclically. Finally, the accuracy results over the k folds is averaged and reported.

The experimental setup of Volansky et al. (2015) is as follows: the dataset is taken from Europarl (Koehn 2005), with approximately 4 million tokens in English and the same number of tokens translated from ten source languages: Danish, Dutch, Finnish, French, German, Greek, Italian, Portuguese, Spanish, and Swedish. The corpus is first tokenized and then partitioned into chunks of approximately 2000 tokens (ending on a sentence boundary). For training and evaluation, each chunk is represented as a *feature vector*. A standard SVM classifier is trained on the feature vectors and is evaluated using ten-fold cross validation. As the task is binary and the dataset is balanced, the baseline accuracy is 50%.

3.2 Features

The crux of the method lies in the selection of features. If the features indeed reflect a true characteristic of translationese, one can assume that a classifier based on these features will be accurate. It should be noted that using content words as features is likely to yield good classifiers: one can expect texts translated, say, from French to English, to include more instances of proper names like *Paris* or even more common nouns like *cheese*. Such features, however, reveal nothing about the properties of translationese, and hence to not advance the goal of the investigation. Volansky et al. (2015) experiment with several feature sets which we briefly list below. Features were selected such that they:

- reflect frequent linguistic characteristics one expects to be present in the two types of text;
- be content-independent, indicating formal and stylistic differences between the texts that are not derived from differences in contents, domain, genre, etc.; and
- be easy to interpret, yielding insights regarding the differences between original and translated texts.

Specifically, features were grouped together to reflect the main translation studies hypotheses. The feature types discussed below have varied dimensionality (some are a single value, averaged over the entire chunk, and some can define very long vectors). If not mentioned otherwise, the value of a feature is a simple count.

Simplification. The simplification hypothesis was modeled through the following features:

Lexical variety. The assumption is that original texts use richer vocabularies than translated ones (Baker 1993; Laviosa 1998). Three different *type-token ratio* (TTR) measures were used, following Grieve (2007), where V is the number of types and N is the number of tokens per chunk. All three versions consider punctuation marks as tokens.

1. V/N, magnified by order of 6.
2. $\log(V)/\log(N)$, magnified by order of 6.
3. $100 \times \log(N)/(1 - V_1/V)$, where V_1 is the number of types occurring only once in the chunk.

Mean word length (in characters). The assumption is that translated texts use simpler words, in particular shorter ones.

Syllable ratio. Assuming that simpler words are used in translated texts, one expects fewer syllables per word.

Lexical density. The frequency of tokens that are *not* nouns, adjectives, adverbs or verbs (Laviosa 1998).

Mean sentence length. Splitting sentences is a common strategy in translation, which is also considered a form of simplification. Baker (1993) renders it one of the universal features of simplification.

Mean word rank. The assumption is that less frequent words are used more often in original texts than in translated ones. This is based on the observation of Blum- Kulka and Levenston (1983) that translated texts "make do with less words" and the application of this feature by Laviosa (1998). A theoretical explanation is provided by Halverson (2003): translators use more prototypical language, i.e., they "regress to the mean" (Shlesinger 1989). To compute this, a list of 6000 English most frequent words was used.

Most frequent words. The frequencies of the N most frequent words in the corpus, where $N = 5, 10, 50$.

Explicitation. Several features were used to model the explicitation hypothesis. The first three feature sets were inspired by an example provided by Baker (1993, pp. 243–4), where the clause *The example of Truman was always present in my mind* was translated into Arabic with a fairly long paragraph, which includes the following: *In my mind there was always the example of the American President Harry Truman, who succeeded Franklin Roosevelt....*

Explicit naming. The ratio of personal pronouns to proper nouns.

Single naming. The frequency of proper nouns consisting of a single token, not having an additional proper noun as a neighbor.

Mean multiple naming. The average length (in tokens) of proper noun sequences.

Cohesive markers. Translations are known to excessively use certain *cohesive markers* (Blum-Kulka 1986; Øverås 1998). A list of 40 such markers was used, based on Koppel and Ordan (2011).

Normalization. Normalization was modeled through the following features:

Repetitions. The number of content words that occur more than once in a chunk.

Contractions. The ratio of contracted forms to their counterpart full form(s).

Average PMI. Original texts are expected to use more collocations, and in any case to use them differently than translated texts. This hypothesis is based on Toury (1980) and Kenny (2001), who showed that translations overuse highly associated words. To reflect this, the average PMI (Church and Hanks 1990) of all bigrams in the chunk was used.

Threshold PMI. The number of bigrams with PMI above 0.

Interference. Several features were selected to model the influence of the source language on the translation product:

POS n-grams. The hypothesis is that grammatical structures used in the various source languages interfere with the translations; and that translations have unique grammatical structure. Following Baroni and Bernardini (2006) and Kurokawa et al. (2009), this assumption was modeled by defining as features unigrams, bigrams and trigrams of part-of-speech (POS) tags.

Character n-grams. Unigrams, bigrams and trigrams of characters. This feature was motivated by Popescu (2011); it captures morphological features of the language.

Prefixes and suffixes. Character n-grams are an approximation of morphological structure. In the case of English, the little morphology exhibited by the language is typically manifested as prefixes and suffixes. A more refined variant of the character n-gram feature, therefore, focuses only on prefixes and suffixes.

Contextual function words. This feature is a variant of POS n-grams, where the n-grams can be anchored by specific function words. It is defined as the frequencies in the chunk of consecutive triplets $\langle w_1, w_2, w_3 \rangle$, where at least two of the elements are function words, and at most one is a POS tag.

Positional token frequency. Writers have a relatively limited vocabulary from which to choose words to open or close a sentence, and the choices may be subject to interference (Munday 1998; Gries and Wulff 2012). The value of this feature is the frequency of tokens appearing in the first, second, antepenultimate, penultimate and last positions in a sentence.

Miscellaneous. Finally, a number of features that cannot be naturally associated with any of the above hypotheses, but nevertheless shed light on the nature of translationese, were also defined.

Function words. Replicating the results of Koppel and Ordan (2011), the same list of function words was used.

Pronouns. Pronouns are function words, and Koppel and Ordan (2011) reported that this subset is among the top discriminating features between originals and translations.

Punctuation. Punctuation marks organize the information within sentence boundaries and to a great extent reduce ambiguity; according to the explicitation hypothesis, translated texts are less ambiguous (Blum-Kulka 1986). The following punctuation marks were used: *? ! : ; - ( ) [ ]* ‘ ’ “ ” / , . Following Grieve (2007), three variants of this feature were defined:

1. The normalized frequency of each punctuation mark in the chunk.
2. A non-normalized notion of frequency: $n/tokens$, where n is the number of occurrences of a punctuation mark; and *tokens* is the actual (rather than normalized) number of tokens in the chunk.
3. n/p, where p is the total number of punctuations in the chunk; and n as above.

Ratio of passive forms to all verbs. The assumption is that English original texts tend to use the passive form more excessively than translated texts, due to the fact that the passive voice is more frequent in English than in some other languages (cf. Teich (2003) for German-English).

The features defined above are all stylistic features that abstract away from the actual contents of the text. As a "sanity check", two content-bearing features were used: token unigrams and token bigrams. These features are expected to yield excellent classifiers but not shed any interesting light on translation hypotheses.

3.3 Results

Each of the feature types discussed above defines a separate classifier. The accuracy of ten-fold cross-validation evaluation with the various feature sets is reported in Table 1.

As expected, the accuracy of the "sanity" features is perfect. This is not surprising in light of their ability to reflect contents, which is highly related to the source language and culture. However, this provides no interesting insights on the properties of translationese.

In contrast, the simplification features tell a mixed story. Some of them are reasonably accurate, especially considering the low dimensionality (of, e.g., TTR). Both TTR and mean word rank provide reasonable separation between the two classes. The other features are less discriminating. Most surprising is mean sentence length which, while providing a better-than-baseline classifier, actually behaves conversely to the hypothesis: as it turns out, the mean sentence length of translations in our corpus is actually *higher* than that of originals (Fig. 1).

The explicitation classifiers also yield mixed results. While the various naming features perform almost at chance level, the cohesive markers turn out to be very effective. In contrast, the normalization features do not discriminate well between originals and translations. In particular, the PMI features again behave conversely to the prediction of the hypothesis: English originals turn out to have much more highly collocated bigrams than translations (Fig. 2).

Table 1. Classification results

Category	Feature	Accuracy (%)
Sanity	Token unigrams	100
	Token bigrams	100
Simplification	TTR (1)	72
	TTR (2)	72
	TTR (3)	76
	Mean word length	66
	Syllable ratio	61
	Lexical density	53
	Mean sentence length	65
	Mean word rank (1)	69
	Mean word rank (2)	77
	N most frequent words	64
Explicitation	Explicit naming	58
	Single naming	56
	Mean multiple naming	54
	Cohesive markers	81
Normalization	Repetitions	55
	Contractions	50
	Average PMI	52
	Threshold PMI	66
Interference	POS unigrams	90
	POS bigrams	97
	POS trigrams	98
	Character unigrams	85
	Character bigrams	98
	Character trigrams	100
	Prefixes and suffixes	80
	Contextual function words	100
	Positional token frequency	97
Miscellaneous	Function words	96
	Pronouns	77
	Punctuation (1)	81
	Punctuation (2)	85
	Punctuation (3)	80
	Ratio of passive forms to all verbs	65

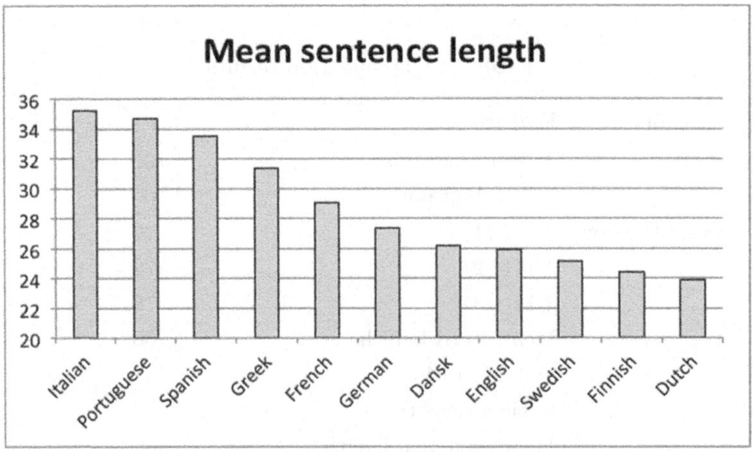

Fig. 1. Mean sentence length of translations from several languages to English vs. original English

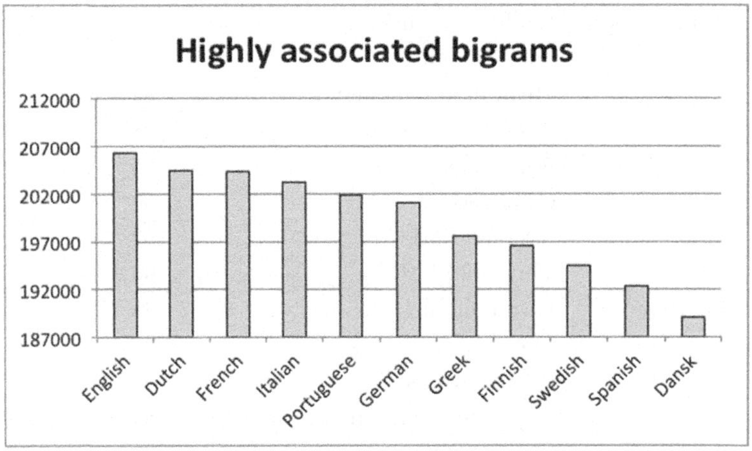

Fig. 2. Number of bigrams whose PMI is above threshold, by source language

Finally, the interference features are clearly the best discriminators. Part-of-speech n-grams, character n-grams, contextual function words and positional token frequency, all of them features that are highly influenced by the structure (and lexis) of the source language, yield excellent, sometimes even perfect classifiers. This result is robust, and persists even when the dimensionality of the feature vectors is reduced (by limiting vectors to the 300 most frequent features) and when experimenting with originals and translations in languages other than English, including Hebrew (Avner et al. 2016), German, and French (Rabinovich and Wintner 2015).

3.4 Unsupervised Classification

Clearly, then, it is possible to automatically distinguish between original and translated texts, with very high accuracy, by employing text classification methods. However, the approaches we surveyed all employed *supervised* machine-learning; they therefore suffer from two main drawbacks: they inherently depend on data annotated as original vs. translated; and they do not scale up to unseen (related or unrelated) domains.[2] These shortcomings undermine the usability of supervised methods for translationese identification in a typical real-life scenario, where no labelled in-domain data are available.

To overcome these issues, Rabinovich and Wintner (2015) proposed to use *un*supervised machine learning, or *clustering*, as a way to identify translationese. In addition to the Europarl corpus described above, they used the following datasets: (i) the Canadian Hansard, transcripts of the Canadian Parliament; (ii) literary classics written (or translated) mainly in the 19th century; and (iii) transcripts of TED and TEDx talks.

First, they replicated the results of Volansky et al. (2015) on the four datasets, using five of the best-performing features. Accuracy is indeed excellent, as shown in Table 2. The results reflect (supervised) ten-fold cross validation evaluation. In the table, 'EUR' stands for Europarl, 'HAN' for Hansard, and 'LIT' for the literary corpus.

Table 2. In-domain (cross-validation) classification accuracy using various feature sets

Feature/corpus	EUR	HAN	LIT	TED
Function words (FW)	96.3	98.1	97.3	97.7
Character trigrams	98.8	97.1	99.5	100.0
POS trigrams	98.5	97.2	98.7	92.0
Contextual FW	95.2	96.8	94.1	86.3
Cohesive markers	83.6	86.9	78.6	81.8

However, when a classifier is trained on one domain and tested on another, the domain-dependence of the supervised method is revealed. This pattern persists even when a classifier is trained on *two* domains and is tested on the third, as shown in Table 3. (The TED corpus was too small to include.)

As these tables clearly show, while in-domain cross-validation evaluation (the rightmost column) shows excellent accuracy, even when the dataset is a mixture of two domains, the classifiers are limited to the domain(s) they were trained on and do not scale up to other datasets.

To remedy the obstacle of domain-dependence, Rabinovich and Wintner (2015) proposed to use unsupervised *clustering* on the entire dataset.

[2] We use "domain" rather freely henceforth to indicate not only the topic of a corpus but also its modality (written vs. spoken), register, genre, date, etc.

Table 3. Pairwise (left) and leave-one-out (right) cross-domain classification using function words

Train / Test	EUR	HAN	LIT	X-validation
EUR		60.8	56.2	96.3
HAN	59.7		58.7	98.1
LIT	64.3	61.5		97.3

Train / Test	EUR	HAN	LIT	X-validation
EUR + HAN			63.8	94.0
EUR + LIT		64.1		92.9
HAN + LIT	59.8			96.0

More specifically, they employed the KMeans algorithm (Lloyd 1982), using KMeans++ initialization (Arthur and Vassilvitskii 2007) and Principal Component Analysis (Pearson 1901) for dimension reduction. Of course, since the method is unsupervised, the labels of the resulting classes are not known; assuming "gold" labels (that is, judging the class by the majority of the instances in it) the accuracy is surprisingly high, as shown in Table 4.

Table 4. Clustering results using various feature sets

Feature/corpus	EUR	HAN	LIT	TED
FW	88.6	88.9	**78.8**	**87.5**
Char trigrams	72.1	63.8	70.3	78.6
POS trigrams	**96.9**	76.0	70.7	76.1
Contextual FW	92.9	**93.2**	68.2	67.0
Cohesive markers	63.1	81.2	67.1	63.0

Furthermore, Rabinovich and Wintner (2015) suggested a simple yet effective method for determining the correct label of the classes and showed that it was perfectly (100%) accurate on all the datasets they experimented with. As the label can be accurately determined, several classifiers, reflecting different feature sets, can be combined in an ensemble, using voting among classifiers to establish the class of each instance. The results of this ensemble clustering are shown in Table 5, and reveal a fully-unsupervised, highly accurate method for discriminating between originals and translations in a single domain.

Finally, Rabinovich and Wintner (2015) defined two simple methods for clustering in a mixed-domain scenario, *flat* and *hierarchical*. The hierarchical method first clusters a mixture of texts into domains (e.g., using KMeans), and then separates each of the resulting (presumably, domain-coherent) clusters into two sub-clusters, presumably originals and translations. The flat approach assumes that the number of domains, k, is known, and attempts to divide the data set into $2 \times k$ clusters, expecting classification by domains and by translationese status, simultaneously. The results, experimenting with a mixture of two and then three different datasets, are shown in Table 6.

Table 5. Clustering consensus by voting

Method/corpus	EUR	HAN	LIT	TED
FW	88.6	88.9	78.8	87.5
FW				
Char trigrams	91.1	86.2	78.2	**90.9**
POS trigrams				
FW				
POS trigrams	**95.8**	89.8	72.3	86.3
Contextual FW				
FW				
Char trigrams				
POS trigrams	94.1	**91.0**	**79.2**	88.6
Contextual FW				
Cohesive markers				

Table 6. Flat and hierarchical clustering of domain-mix using function words

Method/corpus	EUR HAN	EUR LIT	HAN LIT	EUR HAN LIT
Flat	**92.5**	60.7	77.5	66.8
Two-phase	91.3	**79.4**	**85.3**	**67.5**

Summing up, it is possible to accurately identify translationese even in mixed-domain scenarios, but the accuracy of the classification deteriorates as the number of different domains increases.

4 Applications to Machine Translation

The special properties of translationese have ramifications to NLP applications, and in particular to statistical machine translation (SMT). Until recently, research in SMT was divorced from scholarly work in translation studies. This was changed by a series of works, pioneered by Kurokawa et al. (2009) and further elaborated by Lembersky et al. (2011, 2012a, 2012b, 2013) and Twitto et al. (2015). This section summarizes some of the main results reported in these works.

The standard SMT paradigm (Brown et al. 1990, 1993) is based on the *noisy channel model*, whereby the best translation $\hat{T}$ of a source-language sentence S is a target-language sentence T that maximizes some function combining the *faithfulness* of (T, S) and the *fluency* of T. The standard notation assumes that the task is to translate a *foreign* sentence $F = f_1, \cdots, f_m$ into an *English* sentence $E = e_1, \cdots, e_l$. Thus, the best translation is:

$$\begin{aligned}\hat{E} &= \arg\max_E P(E \mid F) \\ &= \arg\max_E \tfrac{P(F|E)\times P(E)}{P(F)} \\ &= \arg\max_E P(F \mid E) \times P(E)\end{aligned}$$

The noisy channel thus requires two components: a *translation model* and a *language model*:

$$\hat{E} = \arg\max_{E \in English} \underbrace{P(F \mid E)}_{Translation\ model} \times \underbrace{P(E)}_{Language\ model}$$

The language model is responsible for the fluency of the translation outcome; it estimates $P(E)$ from a monolingual E corpus. The translation model is responsible for the faithfulness of the translation, and it estimates $P(F \mid E)$ from a bilingual parallel corpus. In addition, a *decoder* is used to produce the most probable E given F, but it will be ignored here.

4.1 Language Models

As mentioned, language models (LMs) are estimated from monolingual corpora of the target language. The common wisdom in SMT used to be that the larger the corpora, the better the translation quality (Brants and Xu 2009); the research question we discuss here is whether this is indeed the case, and in particular, whether corpora compiled from *translated* texts are better for SMT than those compiled from original texts.

Lembersky et al. (2012b) set out to investigate the fitness of language models compiled from translated texts vs. the fitness of LMs compiled from original texts; and whether these differences carry over to SMT, namely whether language models compiled from translated texts are better for MT than LMs compiled from original texts. The fitness of a language model to a reference corpus is evaluated using *perplexity*: the perplexity PP of a language model LM with respect to a sequence of words $w_1, \ldots, w_N$ is defined in terms of the probability LM assigns to the sequence, as follows:

$$PP(LM, w_1 w_2 \ldots w_N) = \sqrt[N]{\prod_{i=1}^{N} \frac{1}{P_{LM}(w_i \mid w_1 \ldots w_{i-1})}}$$

Note that the lower the perplexity, the better the fitness of the LM to the reference set.

Lembersky et al. (2012b) first trained LMs from Europarl corpora and tested them on Europarl reference translations. They experimented with n-gram language models, where n ranged between 1 and 4, and with translations from four different languages (French, German, Spanish and Italian) to English. In all these cases, the results were the same: LMs compiled from translated texts were consistently better than ones compiled from originals; furthermore, LMs compiled from the same source language as the one from which the references were compiled from were the best. Table 7 demonstrates these results for translations from German to English; the results for the other three source languages are very similar. In this and in subsequent tables, 'O-EN' is original English, 'T-XX' refers to translations to English from some language XX, and 'Mix' is a mixture of translated and original texts.

Table 7. The perplexity of various language models on a reference set of German translated into English

LM	1-gram	2-gram	3-gram	4-gram
Mix	451.50	93.00	69.36	66.47
O-EN	*468.09*	*103.74*	*79.57*	*76.79*
T-DE	**443.14**	**88.48**	**64.99**	**62.07**
T-FR	460.98	99.90	76.23	73.38
T-IT	465.89	102.31	78.50	75.67
T-NL	457.02	97.34	73.54	70.56

These results may be attributed to the contents of the various language models, and in particular to specific named entities in them, as mentioned in Sect. 3.2 above. To control this, Lembersky et al. (2012b) further compiled LMs that abstracted away from the actual words in the corpus. They first replaced proper names by a special symbol; then did the same for nouns; and finally, they replaced all words by their parts of speech. In all these cases, the results remained robust, although the differences among the various LMs decreased. In conclusion, LMs compiled from translations, preferably from the same source language as the references, fit the reference set better than LMs compiled from originals.

In order to test the second hypothesis, Lembersky et al. (2012b) trained SMT systems (Koehn et al. 2007) using various LMs and evaluated their quality on a reference set. As is common in SMT, the quality was measured in terms of BLEU scores (Papineni et al. 2002), where the higher the score the better the system is. The results are reported in Table 8. Each pair of columns refers to a specific SMT system, translating from some language into English. The rows indicate the corpus used for the language model.

Obviously, in all cases the best SMT systems are those that use LMs compiled from corpora that were translated from the source language. LMs compiled from other translated texts (from different languages) come next. The worst LMs are those that were compiled from originals. The automated evaluation results were corroborated by manual evaluation, in which humans were asked to assess the quality of the translations. Again, human evaluators preferred the SMT outputs that were produced with the LMs that were based on translations.

Table 8. Quality of SMT using various language models

DE to EN		FR to EN		IT to EN		NL to EN	
LM	BLEU	LM	BLEU	LM	BLEU	LM	BLEU
MIX	21.43	MIX	28.67	MIX	25.41	MIX	24.20
O-EN	*21.10*	O-EN	*27.98*	O-EN	24.69	O-EN	*23.40*
T-DE	**21.90**	T-DE	28.01	T-DE	*24.62*	T-DE	24.26
T-FR	21.16	T-FR	**29.14**	T-FR	25.37	T-FR	23.56
T-IT	21.29	T-IT	28.75	T-IT	**25.96**	T-IT	23.87
T-NL	21.20	T-NL	28.11	T-NL	24.77	T-NL	**24.52**

Going back now to the issue of the size of the training corpus, Fig. 3 plots the quality of French-to-English SMT systems, in terms of BLEU scores, against the size of the French monolingual corpus that was used to train the language model of the systems. The three graphs correspond to a LM compiled from French-translated-to-English texts (FO), a LM compiled from English original texts (EO) and one compiled from a very large corpus which may include originals or translations, in a different domain (Gigaword). Evidently, to reach the same quality obtained using the translated LM, an order of magnitude more original text is needed. The out-of-domain corpus fares much worse. Figure 4 shows a similar plot, where *all* three corpora are out-of-domain, as the reference set used for the evaluation consists of Europarl sentences whereas the LMs are compiled from Hansard or Gigaword materials. Again, much more original data are needed to match the quality of SMT systems built with translated LMs.

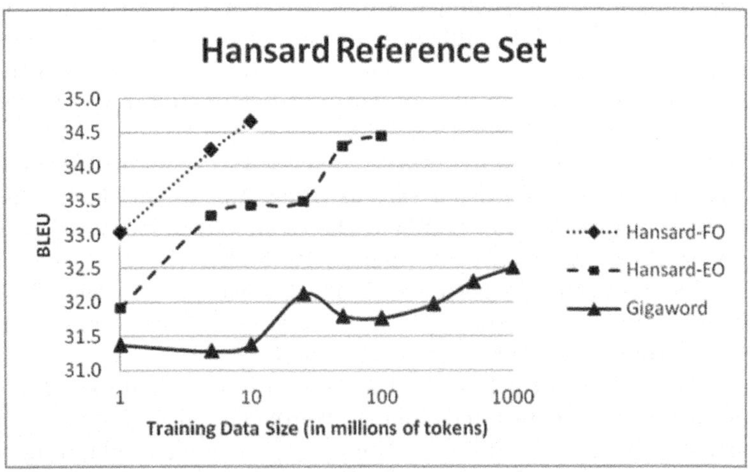

Fig. 3. SMT quality as a function of the size of the language model

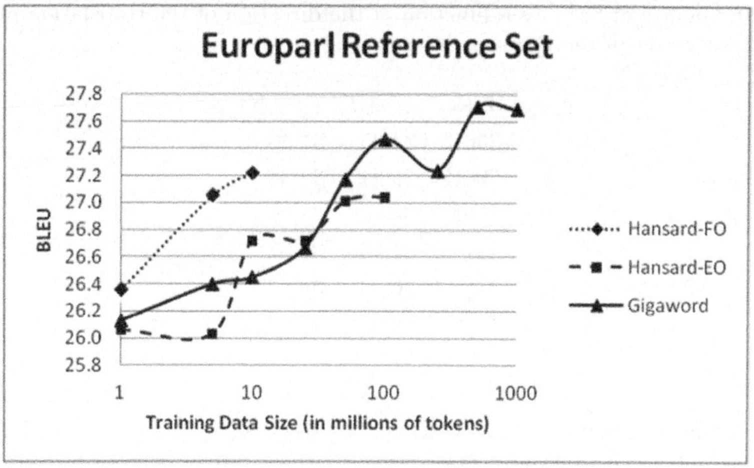

Fig. 4. SMT quality as a function of the size of the language model, out-of-domain evaluation

4.2 Translation Models

In terms of the *translation* model, Kurokawa et al. (2009) have shown that translation models (TMs) compiled from parallel corpora that were (manually) translated in the same direction as that of the SMT task are better than ones translated in the reverse direction. Lembersky et al. (2013) replicated these results, using more varied datasets (both Europarl and the Canadian Hansards) and more language pairs. Table 9 shows the quality of SMT between three languages (six language pairs in total), using TMs compiled in the same direction as that of the task (source-to-target, $S \rightarrow T$) and in the reverse direction (target-to-source, $T \rightarrow S$).

Table 9. Quality of SMT as a function of the direction of the translation of the TM

Task	$S \rightarrow T$	$T \rightarrow S$
FR-EN	33.64	30.88
EN-FR	32.11	30.35
DE-EN	26.53	23.67
EN-DE	16.96	16.17
IT-EN	28.70	26.84
EN-IT	23.81	21.28

Focusing now on a single language pair, namely French to English, Table 10 shows the massive savings in training materials that is facilitated by using $S \rightarrow T$ parallel corpora rather than $T \rightarrow S$ ones. Other language pairs showed very similar patterns.

Table 10. Quality of SMT as a function of the direction of the translation of the TM and of various sizes of the training dataset

Task	$S \rightarrow T$	$T \rightarrow S$
250K	34.35	31.33
500K	35.21	32.38
750K	36.12	32.90
1M	35.73	33.07
1.25M	36.24	33.23
1.5M	36.43	33.73

In a realistic scenario, however, one has access to a large parallel corpus, parts of which were manually translated in the "right" direction, and parts of which in the reverse direction. Lembersky et al. (2013) proposed several methods for adapting the "wrong" subset of the parallel corpus to translationese. The technical details are too complicated to describe here, but the results robustly showed that the SMT systems that resulted from the best adaptation of the translation model were significantly better than ones that used either the entire corpus (in a naïve way) or only its $S \rightarrow T$ subset.

Of course, to benefit from these results, the parallel corpus has to be annotated with information pertaining to the translation direction; such annotation is typically not available. However, Twitto et al. (2015) showed that this obstacle can be overcome, as the predictions of translationese classifiers are as good as meta-information. First, when a monolingual corpus in the target language is given, to be used for constructing a language model, predicting the translated portions of the corpus, and using only them for the language model, is as good as using the entire corpus. Second, identifying the portions of a parallel corpus that are translated in the direction of the translation task, and using only them for the translation model, is as good as using the entire corpus. Twitto et al. (2015) presented results from several language pairs and various data sets, indicating that these results were robust and general.

5 Conclusion

We demonstrated above that awareness of translationese can significantly improve the quality of machine translation. Insights drawn from translation studies can also improve other NLP applications. For example, the task of *native language identification* attempts to identify the mother tongue of non-native writers (typically learners) based on texts they composed in a foreign language (Tetreault et al. 2013). The classifier of Tsvetkov et al. (2013) achieved an accuracy of 80–85% on an 11-way classification task (i.e., texts were authored by native speakers of eleven different languages) using several features that were inspired by the translationese features of Volansky et al. (2015).

The reason that native language identification is similar to the identification of translationese has to do with *interference*. In both of these cases, elements of one linguistic system (the source language in the case of translation, and the native language in the case of non-native speakers) interfere with the production of the target language, and can be traced back by the classifier. In fact, interference is so powerful that it overshadows other, more subtle, properties of translationese. In work in progress, we have been able to demonstrate that translations from related languages (e.g., Spanish and Italian) are closer to each other than translations from more distant languages (e.g., German). This interference is so powerful that it is possible to cluster together related languages based only on their translations to English.

The relations between translationese and non-native language have been explored by Rabinovich et al. (2016), who showed clear similarities but also some significant differences between these two language varieties. In the future, we intend to further explore these relations, focusing not only on advanced, highly fluent non-native speakers but also on learners. We believe that better understanding of the linguistic properties of such language varieties are not only interesting in and of themselves, but may help engineer better NLP systems, as we hope to have shown in this chapter.

Acknowledgements. I am grateful to Noam Ordan for his immense help with the research reported here. Thanks are due to all my other collaborators on these works, including Gennadi Lembersky, Vered Volansky, Udi Avner, Naama Twitto and Ella Rabinovich. Special thanks are due to Agata Savary, not least for her continuous encouragement. I am grateful to the three anonymous reviewers whose constructive comments greatly improved the quality of the presentation. This research was supported by a grant from the Israeli Ministry of Science and Technology.

References

Al-Shabab, O.S.: Interpretation and the Language of Translation: Creativity and Conventions in Translation. Janus, Edinburgh (1996)

Arthur, D., Vassilvitskii, S.: K-means++: the advantages of careful seeding. In: Proceedings of the Eighteenth Annual ACM-SIAM Symposium on Discrete Algorithms, Philadelphia, PA, USA. Society for Industrial and Applied Mathematics, pp. 1027–1035 (2007). http://dl.acm.org/citation.cfm?id=1283383.1283494. ISBN 978-0-898716-24-5

Avner, E.A., Ordan, N., Wintner, S.: Identifying translationese at the word and sub-word level. Digit. Scholarsh. Humanit. **31**(1), 30–54 (2016). http://dx.doi.org/10.1093/llc/fqu047

Baker, M.: Corpus linguistics and translation studies: implications and applications. In: Baker, M., Francis, G., Tognini-Bonelli, E. (eds.) Text and Technology: In Honour of John Sinclair, pp. 233–252. John Benjamins, Amsterdam (1993)

Baroni, M., Bernardini, S.: A new approach to the study of translationese: machine-learning the difference between original and translated text. Lit. Linguist. Comput. **21**(3), 259–274 (2006). http://llc.oxfordjournals.org/cgi/content/short/21/3/259?rss=1

Ben-Ari, N.: The ambivalent case of repetitions in literary translation. Avoiding repetitions: a "universal" of translation? Meta **43**(1), 68–78 (1998)

Blum-Kulka, S.: Shifts of cohesion and coherence in translation. In: House, J., Blum-Kulka, S. (eds.) Interlingual and Intercultural Communication Discourse and Cognition in Translation and Second Language Acquisition Studies, vol. 35, pp. 17–35. Gunter Narr Verlag, Tübingen (1986)

Blum-Kulka, S., Levenston, E.A.: Universals of lexical simplification. Lang. Learn. **28**(2), 399–416 (1978)

Blum-Kulka, S., Levenston, E.A.: Universals of lexical simplification. In: Faerch, C., Kasper, G. (eds.) Strategies in Interlanguage Communication, pp. 119–139. Longman, Harlow (1983)

Brants, T., Xu, P.: Distributed language models. In: Proceedings of Human Language Technologies: The 2009 Annual Conference of the North American Chapter of the Association for Computational Linguistics, Companion Volume: Tutorial Abstracts, Boulder, Colorado, May 2009, pp. 3–4. Association for Computational Linguistics (2009). http://www.aclweb.org/anthology/N/N09/N09-4002

Brown, P.F., Cocke, J., Della Pietra, S.A., Della Pietra, V.J., Jelinek, F., Lafferty, J.D., Mercer, R.L., Roossin, P.S.: A statistical approach to machine translation. Comput. Linguist. **16**(2), 79–85 (1990). ISSN 0891-2017

Brown, P.F., Della Pietra, V.J., Della Pietra, S.A., Mercer, R.L.: The mathematics of statistical machine translation: parameter estimation. Comput. Linguist. **19**(2), 263–311 (1993). ISSN 0891-2017

Church, K.W., Hanks, P.: Word association norms, mutual information, and lexicography. Comput. Linguist. **16**(1), 22–29 (1990). ISSN 0891-2017

Frawley, W.: Prolegomenon to a theory of translation. In: Frawley, W. (ed.) Translation. Literary, Linguistic and Philosophical Perspectives, pp. 159–175. University of Delaware Press, Newark (1984)

Gellerstam, M.: Translationese in Swedish novels translated from English. In: Wollin, L., Lindquist, H. (eds.) Translation Studies in Scandinavia, pp. 88–95. CWK Gleerup, Lund (1986)

Gries, S.T., Wulff, S.: Regression analysis in translation studies. In: Oakes, M.P., Ji, M. (eds.) Quantitative Methods in Corpus-Based Translation Studies. Studies in Corpus Linguistics, vol. 51, pp. 35–52. John Benjamins, Philadelphia (2012)

Grieve, J.: Quantitative authorship attribution: an evaluation of techniques. Lit. Linguis. Comput. **22**(3), 251–270 (2007)

Halverson, S.: The cognitive basis of translation universals. Target **15**(2), 197–241 (2003)

Ilisei, I., Inkpen, D.: Translationese traits in Romanian newspapers: a machine learning approach. Int. J. Comput. Linguist. Appl. **2**(1–2) (2011)

Ilisei, I., Inkpen, D., Corpas Pastor, G., Mitkov, R.: Identification of translationese: a machine learning approach. In: Gelbukh, A. (ed.) CICLing 2010. LNCS, vol. 6008, pp. 503–511. Springer, Heidelberg (2010). doi:10.1007/978-3-642-12116-6_43. http://dx.doi.org/10.1007/978-3-642-12116-6. ISBN 978-3-642-12115-9

Kenny, D.: Lexis and Creativity in Translation: A Corpus-Based Study. St. Jerome, Northampton (2001). ISBN 9781900650397

Koehn, P.: Europarl: a parallel corpus for statistical machine translation. In: Proceedings of the Tenth Machine Translation Summit, pp. 79–86. AAMT (2005). http://mt-archive.info/MTS-2005-Koehn.pdf

Koehn, P., Hoang, H., Birch, A., Callison-Burch, C., Federico, M., Bertoldi, N., Cowan, B., Shen, W., Moran, C., Zens, R., Dyer, C., Bojar, O., Constantin, A., Herbst, E.: Moses: open source toolkit for statistical machine translation. In: Proceedings of the 45th Annual Meeting of the Association for Computational Linguistics Companion Volume Proceedings of the Demo and Poster Sessions, Prague, Czech Republic, pp. 177–180. Association for Computational Linguistics, June 2007. http://www.aclweb.org/anthology/P07-2045

Koppel, M., Ordan, N.: Translationese and its dialects. In: Proceedings of the 49th Annual Meeting of the Association for Computational Linguistics: Human Language Technologies, Portland, Oregon, USA, pp. 1318–1326. Association for Computational Linguistics, June 2011. http://www.aclweb.org/anthology/P11-1132

Kurokawa, D., Goutte, C., Isabelle, P.: Automatic detection of translated text and its impact on machine translation. In: Proceedings of MT-Summit XII, pp. 81–88 (2009)

Laviosa, S.: Core patterns of lexical use in a comparable corpus of English lexical prose. Meta **43**(4), 557–570 (1998)

Laviosa, S.: Corpus-Based Translation Studies: Theory, Findings, Applications. Approaches to Translation Studies. Rodopi, Amsterdam (2002). ISBN 9789042014879

Lembersky, G., Ordan, N., Wintner, S.: Language models for machine translation: original vs. translated texts. In: Proceedings of the 2011 Conference on Empirical Methods in Natural Language Processing, Edinburgh, Scotland, UK, pp. 363–374. Association for Computational Linguistics, July 2011. http://www.aclweb.org/anthology/D11-1034

Lembersky, G., Ordan, N., Wintner, S.: Adapting translation models to translationese improves SMT. In: Proceedings of the 13th Conference of the European Chapter of the Association for Computational Linguistics, Avignon, France, pp. 255–265. Association for Computational Linguistics, April 2012a. http://www.aclweb.org/anthology/E12-1026

Lembersky, G., Ordan, N., Wintner, S.: Language models for machine translation: original vs. translated texts. Comput. Linguist. **38**(4), 799–825 (2012b). http://dx.doi.org/10.1162/COLI_a_00111

Lembersky, G., Ordan, N., Wintner, S.: Improving statistical machine translation by adapting translation models to translationese. Comput. Linguist. **39**(4), 999–1023 (2013). http://dx.doi.org/10.1162/COLI_a_00159

Lloyd, S.: Least squares quantization in PCM. IEEE Trans. Inf. Theory **28**(2), 129–137 (1982). doi:10.1109/TIT.1982.1056489. ISSN 0018-9448

Munday, J.: A computer-assisted approach to the analysis of translation shifts. Meta **43**(4), 542–556 (1998)

Olohan, M.: How frequent are the contractions? A study of contracted forms in the translational English corpus. Target **15**(1), 59–89 (2003)

Øverås, L.: In search of the third code: an investigation of norms in literary translation. Meta **43**(4), 557–570 (1998)

Papineni, K., Roukos, S., Ward, T., Zhu, W.-J.: BLEU: a method for automatic evaluation of machine translation. In: Proceedings of the 40th Annual Meeting on Association for Computational Linguistics, ACL 2002, Morristown, NJ, USA, pp. 311–318. Association for Computational Linguistics (2002). http://dx.doi.org/10.3115/1073083.1073135

Pearson, K.: On lines and planes of closest fit to systems of points in space. Philos. Mag. **2**(6), 559–572 (1901)

Popescu, M.: Studying translationese at the character level. In: Angelova, G., Bontcheva, K., Mitkov, R., Nicolov, N. (eds.) Proceedings of RANLP 2011, pp. 634–639 (2011)

Rabinovich, E., Wintner, S.: Unsupervised identification of translationese. Trans. Assoc. Comput. Linguist. **3**, 419–432 (2015). https://tacl2013.cs.columbia.edu/ojs/index.php/tacl/article/view/618. ISSN 2307-87X

Rabinovich, E., Nisioi, S., Ordan, N., Wintner, S.: On the similarities between native, non-native and translated texts. In: Proceedings of the 54th Annual Meeting of the Association for Computational Linguistics, ACL 2016, pp. 1870–1881, August 2016. http://aclweb.org/anthology/P/P16/P16-1176.pdf

Shlesinger, M.: Simultaneous interpretation as a factor in effecting shifts in the position of texts on the oral-literate continuum. Master's thesis, Tel Aviv University, Faculty of the Humanities, Department of Poetics and Comparative Literature (1989)

Teich, E.: Cross-Linguistic Variation in System and Text: A Methodology for the Investigation of Translations and Comparable Texts. Mouton de Gruyter, Mouton (2003)

Tetreault, J., Blanchard, D., Cahill, A.: A report on the first native language identification shared task. In: Proceedings of the Eighth Workshop on Building Educational Applications Using NLP. Association for Computational Linguistics, June 2013

Toury, G.: In Search of a Theory of Translation. The Porter Institute for Poetics and Semiotics, Tel Aviv University, Tel Aviv (1980)

Toury, G.: Descriptive Translation Studies and Beyond. John Benjamins, Amsterdam/Philadelphia (1995)

Tsvetkov, Y., Twitto, N., Schneider, N., Ordan, N., Faruqui, M., Chahuneau, V., Wintner, S., Dyer, C.: Identifying the L1 of non-native writers: the CMU-Haifa system. In: Proceedings of the Eighth Workshop on Innovative Use of NLP for Building Educational Applications, pp. 279–287. Association for Computational Linguistics, June 2013. http://www.aclweb.org/anthology/W13-1736

Twitto, N., Ordan, N., Wintner, S.: Statistical machine translation with automatic identification of translationese. In: Proceedings of the Tenth Workshop on Statistical Machine Translation, Lisbon, Portugal, pp. 47–57. Association for Computational Linguistics, September 2015. http://aclweb.org/anthology/W15-3002

van Halteren, H.: Source language markers in EUROPARL translations. In: Scott, D., Uszkoreit, H., (eds.) Proceedings of the 22nd International Conference on Computational Linguistics, COLING 2008, Morristown, NJ, USA, pp. 937–944. Association for Computational Linguistics (2008). ISBN 978-1-905593-44-6

Vanderauwerea, R.: Dutch Novels Translated into English: The Transformation of a 'Minority' Literature. Rodopi, Amsterdam (1985)

Volansky, V., Ordan, N., Wintner, S.: On the features of translationese. Digit. Scholarsh. Humanit. **30**(1), 98–118 (2015)

Two Decades of Pattern Mining: Principles and Methods

Arnaud Soulet(✉)

University of Tours, Blois, France
arnaud.soulet@univ-tours.fr

Abstract. In 1993, Rakesh Agrawal, Tomasz Imielinski and Arun N. Swami published one of the founding papers of pattern mining: "Mining Association Rules Between Sets of Items in Large Databases". It aimed at enumerating the complete collection of regularities observed in a given dataset like for instance sets of products purchased together in a supermarket. For two decades, pattern mining has been one of the most active fields in Knowledge Discovery in Databases. This paper presents an overview of pattern mining. We first present the principles of language and interestingness that are two key dimensions for defining a pattern mining process to suit a specific task and a specific dataset. The language defines which patterns can be enumerated (itemsets, sequences, graphs). The interestingness measure defines the archetype of patterns to mine (regularities, contrasts or anomalies). Starting from a language and an interestingness measure, we depict the two main categories of pattern mining methods: enumerating all the patterns whose interestingness exceeds a user-specified threshold (satisfaction problem) or enumerating all the patterns whose interest is maximum (optimization problem). Finally, we present an overview of interactive pattern mining which aims at discovering the user's interest while mining relevant patterns.

Keywords: Data mining · Pattern mining

1 Introduction

In 1993, Rakesh Agrawal, Tomasz Imielinski and Arun N. Swami published one of the seminal papers of pattern mining [1]: "Mining Association Rules between Sets of Items in Large Databases" in the proceedings of the ACM SIGMOD International Conference on Management of Data by introducing the problem of extracting interesting association rules. Formally, this problem is to enumerate all the rules of type $X \rightarrow I$ where X is a set of items and I is an item not found in X such that the probabilities $P(X, I)$ and $P(I|X)$, respectively estimated by *support* and *confidence*, are sufficiently high. This seminal paper has initiated a school of thought strongly influenced by the field of databases. In contrast to the field of Machine Learning, a particular attention is paid to sound and complete extractions while the evaluation is mainly based on the speed and the amount of memory required. A recent bibliometric survey [2] analyzed the work

P. Marcel and E. Zimányi (Eds.): eBISS 2016, LNBIP 280, pp. 59–78, 2017.
DOI: 10.1007/978-3-319-61164-8_3

related to pattern discovery published from 1995 to 2012 based on more than one thousand papers (1,087 papers devoted to pattern mining from the 6,888 papers published in the five major conferences on Knowledge Discovery in Databases: KDD, PKDD, PAKDD, ICDM and SDM). This study shows that pattern mining is an important subfield of Knowledge Discovery in Databases since about one paper out of six concerns the former. About 20% of authors from these five conferences have contributed to at least one publication in pattern mining.

Pattern mining is based on two key dimensions that each new proposal must consider: the language and the interestingness. Basically, the language defined the syntax of mined patterns while the interestingness measure defined the semantics of mined patterns.

- **Language:** Language is the domain of definition of patterns that are enumerated. While most methods consider association rules and itemsets as the pattern language at hand, in the past decade a clear trend towards more sophisticated representations has emerged. The community effort in terms of number of published papers focuses on the most complex languages such as sequences or sub-graphs. This variability of language allows pattern mining to address highly structured data without flattening them. Like works in Artificial Intelligence, a specialization relation on this language makes the learning of concepts by induction possible [3]. This specialization relation determines whether a pattern is observed or not in an entry of the dataset.
- **Interestingness:** Once the language and its specialization relation are defined, it remains to define what are the interesting patterns. In most cases the interestingness of a pattern is evaluated by a measure. For instance, the frequency of a pattern (i.e., the number of occurrences of the pattern within the dataset) is often used to judge the importance of a pattern. Intuitively, the basic idea is to consider that a pattern which occurs in many data observations is interesting. However, this measure does not cover all possible semantics (e.g., contrast or exceptional patterns) and the frequency tends to favor spurious patterns. These two obstacles have motivated a large number of works on interestingness measures.

In this context, the general idea of a pattern mining process is to choose the right language and the right interestingness measure according to the task and the dataset and then, to apply a mining method. These mining methods are mainly divided into two categories:

- Constraint-based pattern mining [4,5] aims at extracting all the patterns that satisfy a Boolean predicate as interestingness criterion. Most often this predicate, called constraint, requires that an interestingness measure exceeds a given threshold. The challenge is to achieve a sound and complete enumeration despite the huge size of the search space (stemming from the language) and the complexity of the constraint (stemming from the interestingness measure). For this purpose, pruning properties were introduced for different classes of constraints [6–8].

- Preference-based pattern mining [9,10] aims at extracting the most preferred patterns. This notion of preference relies on a binary relation between patterns specified by the user. For instance a pattern will be preferred over another if its value for a measure is higher. The fact of not having to set a threshold (contrary to constraint-based approaches) facilitates the definition of the problem by the user. But, it further complicates the mining step which has to determine the threshold during the scanning of the search space.

Recently, the need of an interest criterion explicitly specified by the user has been questioned. Indeed, it is often difficult for an end user to know in advance which is the right constraint or preference relation modeling their interest. In practice, the adjustment of measures and thresholds in a mining process quickly becomes tedious. Rather than asking the user to explicitly define their interest, interactive pattern mining [11] captures it based on their user feedback about some preliminary patterns. This promising approach, however, raises issues on setting and learning the user preference model. Moreover, the interaction requires that the extraction of patterns is instantaneous [12] in order to have a tight coupling between the user and the mining system.

This paper is a brief introduction to pattern mining that benefits from the formal framework introduced in [4]. It unfortunately leaves out many works like pattern set mining [13] or declarative frameworks [14]. The list of methods presented here is not exhaustive and no algorithm is given due to lack of space. This paper is intended as an entry point and many references link to deeper survey papers. It introduces the main aspects of pattern mining, major issues and the methods' principles. Finally, a bibliometric analysis describes a few trends about languages (see Sect. 2.2) and interestingness measures (see Sect. 3.2) based on [2]. This study focuses on the proceedings of all the conferences whose title contains "data mining" and are ranked A by The Computing Research and Education Association of Australia[1]: KDD (ACM SIGKDD International Conference on Knowledge Discovery and Data Mining[2]), PKDD[3] (European Conference on Principles of Data Mining and Knowledge Discovery[4]), PAKDD (Pacific Asia Knowledge Discovery and Data Mining[5]), ICDM (IEEE International Conference on Data Mining[6]) and SDM (SIAM International Conference on Data Mining[7]).

Section 2 introduces the different notations concerning the language while Sect. 3 outlines the main aspects of interestingness measures. Then, Sect. 4 introduces the problem of constraint-based pattern mining and the main principles

[1] www.core.edu.au, 2010.

[2] www.kdd.org.

[3] PKDD was attached in 2001 to ECML (European Conference on Machine Learning) then the two conferences merged in 2008. Since 2008, PKDD corresponds to ECML/PKDD.

[4] www.ecmlpkdd.org.

[5] www.pakdd.org.

[6] www.cs.uvm.edu/~icdm.

[7] www.siam.org/meetings/archives.php#sdm.

of extraction methods. Section 5 follows the same schema with preference-based pattern mining. Interactive pattern mining is described in Sect. 6 where a general framework is given and instant pattern mining is introduced. Finally, Sect. 7 concludes this paper.

2 Pattern, Language and Dataset

2.1 Basic Definitions

Pattern mining has the advantage of processing highly structured data while most Data Mining approaches are only dedicated to flat data (a collection of records as attribute-value)[8]. This structured data is a *language* $\mathcal{L}$ and a dataset $\mathcal{D}$ is a multiset of $\mathcal{L}$ [9]. Table 1(a) presents such a dataset $\mathcal{D}$ gathering 5 movies where Leonardo DiCaprio plays approximately the same character[10]. There are 5 movies $m_1, \ldots, m_5$ described by 4 items: Troubled romantic (denoted by T), Rich (R), Dies (D) and Hidding Secret (H). For instance, the first transaction describes the movie "Titanic" where Leonardo DiCaprio plays a character that dies.

Table 1. Toy datasets for itemsets and sequences

$\mathcal{D}$

Movie	Title	Troubled romantic	Rich	Dies	Hidding Secret
m_1	Titanic	T		D	
m_2	Catch Me If You Can	T			H
m_3	Inception	T	R	D	
m_4	Django Unchained		R	D	
m_5	The Great Gatsby	T	R	D	H

(a) Itemset dataset

$\mathcal{D}_{seq}$

Trans	Sequences
t_1	$\langle(A)(C)\rangle$
t_2	$\langle(AD)\rangle$
t_3	$\langle(AB)(AC)\rangle$
t_4	$\langle(B)(A)(CD)\rangle$
t_5	$\langle(C)(C)(D)\rangle$

(b) Sequential dataset

Pattern mining is learning by induction method consisting in finding the patterns in $\mathcal{L}$ that correctly generalize the transactions $\mathcal{D}$. For this purpose, we use a *specialization relation* $\preceq$ which is a partial order relation on $\mathcal{L}$ [4] such that a pattern φ covers more transactions than a more specific pattern γ: $\varphi \preceq \gamma \wedge \gamma \preceq t \Rightarrow \varphi \preceq t$ for all $\varphi \in \mathcal{L}$, $\gamma \in \mathcal{L}$ and $t \in \mathcal{L}$. When $\varphi \preceq \gamma$, we say both that φ is more general than γ, or γ is more specific than φ. For instance, the itemset $\{T\}$ is more general than the itemset $\{T, R\}$ w.r.t $\subseteq$ (i.e., the set inclusion is a specialization relation for itemsets). As $\{T\}$ is more general

[8] Conversely, pattern mining is not well adapted to continuous data even if there exist proposals [15,16].

[9] In this paper, we use the same language for mined patterns and the dataset. It is nevertheless possible to use distinct languages benefiting from a cover relation.

[10] This toy example is inspired from an article found on the Web site mic.com.

than {T, R}, it is sure that {T} covers more transactions in $\mathcal{D}$ than {T, R} in Table 1(a). In the following, for simplicity, most of the examples are illustrated with itemsets that are representing as strings: T and TR respectively mean {T} and {T, R}. Note that association rules can also be derived from itemsets. The association rule T → R says that when DiCaprio plays a Troubled romantic (T) character, he is rich (R).

Table 1(b) also illustrates this framework by providing a sequential dataset [17,18]. For instance, the transaction $t_4 = \langle(B)(A)(CD)\rangle$ represents a first event B, followed by an event A, followed by the conjunction of events C and D. In this context, the sequential pattern $\langle(B)(C)\rangle$ is more general than $\langle(A\mathbf{B})(A\mathbf{C})\rangle$ or $\langle(\mathbf{B})(A)(\mathbf{C}D)\rangle$ given that for two sequential patterns $\varphi = \langle(X_1)\dots(X_n)\rangle$ and $\gamma = \langle(Y_1)\dots(Y_m)\rangle$, φ is more general than γ, denoted by $\varphi \sqsubseteq \gamma$, iff there exist n indexes $i_1 < \dots < i_n$ such that $X_j \subseteq Y_{i_j}$ for $1 \leq j \leq n$.

It is clear that the choice of language $\mathcal{L}$ and specialization relation $\preceq$ defines the structure of the discovered patterns and that choice is as important as it is difficult. For instance, for Text Mining, it is possible to choose different languages: representation with bags of words (itemsets) or by considering an order on the words (sequences). Assuming a sequential representation, it is possible to generalize sequences with gaps (as proposed above with $\sqsubseteq$) or otherwise, sequences without gaps (by adding a constraint of adjacency on indexes). Of course, choosing a language over an other one will impact the mined patterns and consequently, the analysis that results.

Beyond the knowledge representation, the language raises several important challenges to the mining methods described in next sections (Sects. 4–6). The first challenge is to curb the combinatorial explosion. With only 3 items and a length of 3, it is possible to build 8 itemsets, 80 sequential patterns, 238 subgraphs. We will see that the methods rely on pruning techniques that exploit the properties of interestingness criteria. For example, an anti-monotone property of constraints reduces the search space (see Sect. 4.2). The second challenge is the exploration of language without redundancy to avoid enumerating multiple times the same pattern. For itemsets, using lexicographical order avoids considering twice the same itemset. For more complex languages, it is necessary to use canonical forms. The last challenge is to compare patterns to each other to implement a specialization relation [19]. While it does not pose any difficulty for itemsets, graph comparison raises isomorphism problems. Again, a canonical form often facilitates isomorphism tests.

At this stage, the pattern mining problem can be formulated as follows: **Given a pattern language $\mathcal{L}$ and a dataset $\mathcal{D}$, find interesting patterns of $\mathcal{L}$ present in $\mathcal{D}$.** This formulation of this problem nonetheless conceals a crucial issue: what is an interesting pattern? For instance, are the rules T → R and T → D interesting? The answer to these questions is addressed in Sect. 3. But before, the section below examines the prevalence of languages having different complexities.

2.2 Language Sophistication

In Table 2, 1,087 papers concerning pattern mining are shown sorted into 7 different categories. Note that "itemset" category also includes association rules which are often derived from itemsets. "generic" means that the proposal of the paper works at the same time for different languages.

Table 2. List of languages

Language	Number	Proportion
itemset (association rules, sets)	685	0.64
sequence (episode, string, periodic, temporal)	190	0.17
graph (molecular, structure, network)	107	0.10
tree (xml)	49	0.05
spatial and spatio-temporal	30	0.03
generic	18	0.02
relational	8	0.01

As expected, association rules and itemsets, which are at the origin of pattern mining, are the most studied with approximately 2/3 of the whole corpus. About a quarter of papers concerns sequences and graphs. The discovery of patterns in spatio-temporal data and relational data remains quite marginal. More surprisingly, we find that very few studies have addressed generic approaches in terms of language. A probable explanation is the difficulty of proposing a general framework both theoretically and in terms of implementation like [4, 7, 20]. Furthermore, Fig. 1 depicts the evolution of the three most representative languages during the past two decades. The plots report the results given in absolute (left) and in percentage (right).

Table 2 shows that the more complex a language, the fewer papers dedicated to it. First, the intrinsic complexity related to the combinatorial problem makes it difficult to exhaustively extract patterns when sophisticated languages are involved (as explained above). Second, the evolution of this sophistication of language was gradual as described in Fig. 1: itemsets, sequences and then, graphs. In fact, the knowledge gained with the first languages has reduced the number of scientific challenges for the next languages. For instance, pruning methods of the search space for itemsets (based on anti-monotonicity for instance) are transferable to other languages.

Nevertheless, we observe one exception with trees which are less studied than graphs. Trees are sometimes simplified to be treated as variants of sequences or as special cases of graphs. While the proportion of publications concerning rules and itemsets decreases, the more sophisticated languages (i.e., sequences and subgraphs) continue to progress in pattern mining (see Fig. 1). A survey [19] confirms the importance of subgraph mining between 1994 and 2007 through

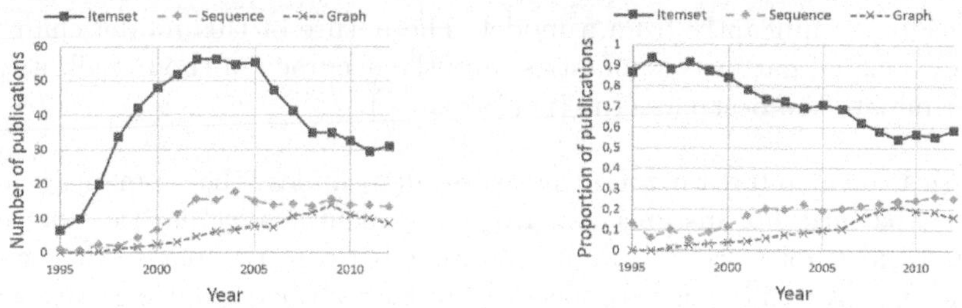

Fig. 1. Evolution of the number of publications per language

bibliometric information. However, this sophistication reaches its limit and no language seems to succeed to graphs because there is not a significant amount of papers about spatio-temporal or relational patterns. These data may not be available in sufficient quantity while those available are reduced to simpler languages such as graphs.

3 Interestingness Measures

3.1 Basic Definitions

Pattern discovery takes advantage of interestingness measures to evaluate the relevancy of a pattern. The *frequency* of a pattern φ in the dataset $\mathcal{D}$ can be considered as the number of transactions covered by φ [21]: $freq(\varphi, \mathcal{D}) = |\{t \in \mathcal{D} \text{ such that } \varphi \preceq t\}|$. A pattern is said to be *frequent* when its frequency exceeds a user-specified minimal threshold. For instance, in Table 1, the pattern T is frequent with 2 as minimal threshold because $freq(\mathrm{T}, \mathcal{D}) = |\{m_1, m_2, m_3, m_5\}|$ (≥ 2). In the same way, the frequency of the association rule T $\rightarrow$ R (resp. T $\rightarrow$ D) is 2 (resp. 3). In his filmography, when DiCaprio plays a troubled romantic character, he dies more often than he is rich.

The frequency was the first used interestingness measure and it is the most popular (see Sect. 3.2). There are a significant number of measures to embrace all useful semantics with varying performance depending on their complexity [22]. A few examples are given in Table 3. In general, a pattern is considered relevant

Table 3. Different examples of interestingness measures

Interestingness measure	Definition
Support	$\frac{\lvert\{t \in \mathcal{D} \text{ such that } \varphi \preceq t\}\rvert}{\lvert\mathcal{D}\rvert}$
Area	$supp(X, \mathcal{D}) \times \lvert X\rvert$
Lift	$\frac{supp(X,\mathcal{D})}{\Pi_{i \in X} supp(i,\mathcal{D})}$
Productivity	$\min_{Y \subset X} \frac{supp(X,\mathcal{D})}{supp(Y,\mathcal{D}) \times supp(X \setminus Y,\mathcal{D})}$
Growth rate	$\frac{supp(X,\mathcal{D}_1)}{supp(X,\mathcal{D}_2)}$

if it deviates significantly from a model. The nature of this model changes the type of extracted patterns (semantics) and the accuracy of the model, its ability to discriminate the best patterns (performance).

Semantics. An interestingness measure determines the semantics of the extracted pattern. For instance, the frequency identifies regularities that appear in data. It does not work for mining contrasts between two parts of the datasets where a frequency difference is expected. For this purpose, contrast measures like the growth rate defined in Table 3 are better suited (where their value increases with the support in $\mathcal{D}_1$ when the support in $\mathcal{D}_2$ remains constant). Similarly, the frequency cannot isolate rare phenomena that are not recurring by definition. In that case, lift or productivity are more interesting because they measure a variation between the true support and the expected one. Rather than just considering the occurrences of the pattern within the dataset, it may be appropriate to consider its utility (e.g., cost or profitability). In the case of association rule mining, the confidence of a rule $X \rightarrow Y$ estimates the probability of Y given X. Interestingly, the confidence of $\mathrm{T} \rightarrow \mathrm{D}$ is $freq(\mathrm{TD}, \mathcal{D})/freq(\mathrm{T}, \mathcal{D}) = 3/4$ is higher than that of $\mathrm{T} \rightarrow \mathrm{R}$ (which is $freq(\mathrm{TR}, \mathcal{D})/freq(\mathrm{T}, \mathcal{D}) = 2/4$) meaning that when DiCaprio plays a troubled romantic character, he is more likely to die than to be rich.

Performance. The performance of an interestingness measure varies with the quality of its underlying model [23]. To find correlations, the data model can be defined from an independence model, a Bayesian model, a maximum entropy model and so on. In that way a general framework is proposed by Exceptional Model Mining [24]. For instance, lift and productivity respectively rely on an independence model on items and on a partition model. Of course, the more accurate the model, the more efficient the measure. In practice, productivity is much more selective than lift and more efficient for isolating the most correlated patterns [25]. In order to illustrate this notion of performance with the association rule $\mathrm{T} \rightarrow \mathrm{D}$, we are going to compare its confidence ($= 3/4$ as seen above) and its lift. The lift of $\mathrm{T} \rightarrow \mathrm{D}$ equals to $supp(\mathrm{TD}, \mathcal{D})/(supp(\mathrm{T}, \mathcal{D}) \times supp(\mathrm{D}, \mathcal{D})) = 0.6/(0.8 \times 0.8) = 0.9375 < 1$. It is therefore a slight negative correlation because the lift is slightly less than 1. Unlike the conclusion drawn with confidence, being troubled romantic does not (fortunately) increases the chances of dying.

Although works [26] have identified properties that a well-behaved measure has to satisfy, capturing the interestingness with a measure remains a complicated issue. Its definition is even more complex than it has to be suited for efficiently enumerating all relevant patterns. For instance, Fig. 2 depicts two lattices of itemsets with a grayscale proportional to the interest. It is easy to observe that the darkest itemsets for frequency (on the left) are concentrated in the top of the lattice while those for area (on the right) are disseminated throughout the lattice. In fact, the frequency is an anti-monotone function meaning that when we consider two patterns $\varphi \preceq \gamma$, the frequency of φ is greater than that of γ (in comparison area has no good property). Therefore it will be algorithmically harder to enumerate interesting patterns according to the area (see Sects. 4 and 5).

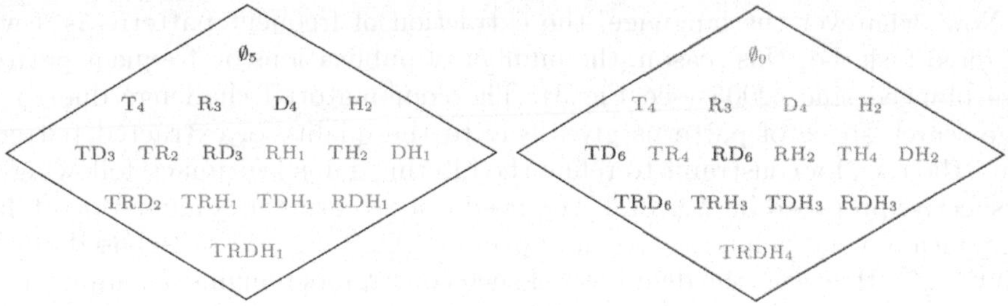

Fig. 2. Lattices of itemsets with a grayscale proportional to interestingness (frequency on the left and area on the right)

Now we can rephrase the pattern mining problem as follows: **Given a pattern language $\mathcal{L}$, a dataset $\mathcal{D}$ and an interestingness measure m, find interesting patterns of $\mathcal{L}$ with respect to m present in $\mathcal{D}$.** This new formulation still contains an ambiguity in the definition of what an interesting pattern is. Constraint-based pattern mining (see Sect. 4) judges a pattern X as interesting for m as soon as $m(X, \mathcal{D})$ is greater than a user-specified threshold (satisfaction problem). Preference-based pattern mining (see Sect. 5) considers that a pattern X is interesting for m when no pattern (or only k patterns) has a better value for m (optimization problem).

3.2 The Obsession with Frequency

This section briefly analyzes the prevalence of different interestingness categories for 538 pattern mining papers (see Table 4). Overall, the minimal frequency constraint with 50% of publications is by far the most used. Indeed, many papers address the frequent pattern mining described above so as to provide a new or more effective algorithm by varying either the language in input or the condensed representation in output (see Sect. 5.2 for a definition of condensed representations).

Table 4. List of interestingness measures

Interestingness	Number	Proportion
Regularity (frequent, support, area)	263	0.48
Significant (chi-square, correlated)	107	0.21
Contrast (emerging, discriminative)	72	0.13
Generic (monotone, anti-monotone, convertibe)	42	0.08
Exception (abnormal, surprising, anomaly, unexpected)	32	0.06
Utility	22	0.04

Now, whatever the language, the extraction of frequent patterns is a well-mastered task. For this reason, the number of publications on frequent patterns have plunged since 2005 (see Fig. 3). The combinatorial challenge due to the large search space of patterns gives way to the quality of extracted patterns. Thus, the use of a constraint to refine the filtering gains legitimacy following the perspective proposed by Agrawal: "we need work to bring in some notion of 'here is my idea of what is interesting,' and pruning the generated rules based on that input." [27]. However, the definition of such constraints remains a complex issue. The proposal of a general theory of *Interestingness* was already indicated as a challenge for the past decade by Fayyad et al. in 2003 [28]. Later, Han et al. [29] follow the same idea: "it is still not clear what kind of patterns will give us satisfactory pattern sets in both compactness and representative quality".

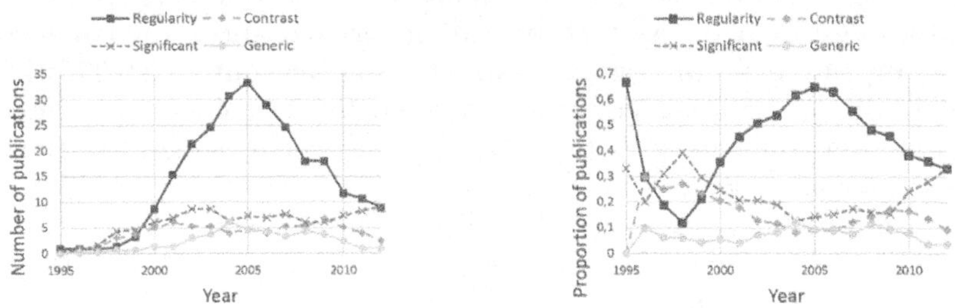

Fig. 3. Evolution of the number of publications per constraint

4 Constraint-Based Pattern Mining

4.1 Principle

A large part of the published literature about pattern mining consists in extracting all the patterns of a language $\mathcal{L}$ that are relevant where the relevance is modeled by a predicate, called *constraint*. Often this predicate selects all patterns whose value for an interestingness measure is greater than a given threshold. For instance, the extraction of frequent patterns enumerates all the patterns whose frequency is greater than a minimum threshold. In general, this task is called *constraint-based pattern mining* [4]:

Problem 1 (Constraint-based pattern mining). Given a language $\mathcal{L}$, a dataset $\mathcal{D}$ and a constraint q, constraint-based pattern mining aims at enumerating all the patterns in $\mathcal{L}$ that satisfy q in $\mathcal{D}$:

$$\mathcal{Th}(\mathcal{L}, \mathcal{D}, q) = \{\varphi \in \mathcal{L} : q(\varphi, \mathcal{D}) \text{ is true}\}$$

This set of patterns is called the theory.

Table 5. Anti-monotone pruning based on frequency in a Breadth-First Search traversal

Pattern	Frequency	Pruned patterns
∅	5	
T	4	
R	3	
D	4	
H	2	
TR	2	
TD	3	
TH	2	
RD	3	
~~RH~~	1	RDH, TRH, TRDH
~~DH~~	1	RDH, TDH, TRDH
TRD	2	

With this framework, frequent itemset mining is formalized as the theory $Th(2^{\mathcal{I}}, \mathcal{D}, freq(\varphi, \mathcal{D}) \geq 2) = \{\emptyset, \mathrm{T}, \mathrm{R}, \mathrm{D}, \mathrm{H}, \mathrm{TR}, \mathrm{TD}, \mathrm{RD}, \mathrm{TH}, \mathrm{TRD}\}$. In practice, the calculation of this theory cannot be done with a naive enumeration of all itemsets belonging to the language because this language is too large a search space (whose size is exponential with the number of items). It is then necessary to apply pruning techniques stemming from the constraint and the language. The principle of these pruning methods relies on the following property:

Property 1 (Safe pruning). Given a candidate pattern set $S \subseteq \mathcal{L}$ such that $Th(\mathcal{L}, \mathcal{D}, q) \subseteq S$, we have: $Th(\mathcal{L}, \mathcal{D}, q) = Th(S, \mathcal{D}, q)$.

The smaller the set of candidates S, the more efficient the extraction. During the enumeration, this set S is built dynamically, the patterns are considered one by one (in breadth-first [21] or depth-first manner [30]) and for each pattern φ, a part of the language may be excluded from S. For instance, in our toy example, RH is not frequent (its frequency is only 1) and then it is sure that all supersets of RH are more specific and then, are not frequent either. Thus, the three patterns TRH, RDH and TRDH are excluded from the candidate pattern set S. Table 5 provides the complete mining with a breadth-first search approach. We note that only two non-frequent patterns are visited (RH and DH). For other more advanced techniques (especially data structures), [31] surveys frequent itemset mining algorithms.

4.2 From Frequency to Better Interestingness Measures

As discussed in Sect. 3, support measure is far from being really interesting for many tasks and more sophisticated measures have been investigated. However, sound and complete mining imposes some limitations on the constraint definition for deriving safe pruning properties. The principle seen above for frequent pattern mining works for all anti-monotone constraints i.e., constraints q such that $(\forall\varphi \preceq \gamma)(q(\gamma) \Rightarrow q(\varphi))$. There are also more complex classes of constraints [6–8,32]. For instance, convertible constraints [7] can be reduced to anti-monotone constraints by enumerating the search space in the right order. We refer the reader to [5] for a deeper discussion about classes of constraints. For very complex constraints where the solution space has exploded throughout the lattice, the idea is to find a relaxed constraint q' such that $(\forall\varphi \in \mathcal{L})(q(\varphi) \Rightarrow q'(\varphi))$ and at the same time, the relaxed constraint q' is anti-monotone [8]. For instance, the constraint $area \equiv \mathit{freq}(\varphi, \mathcal{D}) \times |\varphi| \geq a$ is a complex constraint which is not anti-monotone. But $\mathit{freq}(\varphi, \mathcal{D}) \times 4 \geq a$ is an anti-monotone constraint that implies $area$ (considering the toy dataset where the longest transaction has 4 items). During the past decade, declarative and effective approaches have also been proposed benefiting from Constraint Programming [14].

Limits. Constraint-based pattern mining is an elegant framework but it suffers from several issues. First, it is difficult for an end user to define interestingness by a constraint. In particular, the choice of thresholds (that are "crisp") is not easy (yet critical). Besides, when the user succeeds to define their constraint, this approach often returns a huge number of patterns (even with the most advanced constraints). Sometimes the amount of mined patterns is far beyond the size of the original dataset because the size of the language exponentially grows with the number of items. It is then impossible for the user to explore and analyze this collection of patterns.

5 Preference-Based Pattern Mining

5.1 Principle

As constraint-based pattern mining often returns too many patterns, a lot of proposals are intended to focus on the best patterns according to a user-specified preference order. This preference relation is a binary relation R (partial or total), where $\varphi R \gamma$ means that φ is preferred to γ and, γ is dominated by φ. For example, the pattern φ is preferred to γ if its frequency is higher: $(\varphi R_{\mathit{freq}} \gamma) \Leftrightarrow (\mathit{freq}(\varphi, \mathcal{D}) > \mathit{freq}(\gamma, \mathcal{D}))$. In this paper, this task is called *preference-based pattern mining*, but in the literature, it is also referred as dominance programming [9] or optimal pattern mining [33]:

Problem 2 (Preference-based pattern mining). Given a language $\mathcal{L}$, a dataset $\mathcal{D}$ and a preference relation R, preference-based pattern mining aims at mining all the patterns which are not dominated by at least k patterns:

$$\mathcal{B}est_k(\mathcal{L}, \mathcal{D}, R) = \{\varphi \in \mathcal{L} : \text{there is no } k \text{ patterns } \gamma_i \in \mathcal{L} \text{ such that } \gamma_i R \varphi\}$$

One of the advantages of this approach is that the threshold k is often quite easy to set for a end user. In the case of the extraction of the k best patterns according to an interestingness measure, this threshold corresponds to the number of patterns to be extracted. For instance, the top-3 frequent itemset mining [34] is defined as $\mathcal{B}est_3(2^{\mathcal{I}}, \mathcal{D}, R_{freq}) = \{\emptyset, \mathrm{T}, \mathrm{D}\}$ and only returns 3 itemsets.

For the same reasons as for constraint-based pattern mining, it is not possible to enumerate all the patterns of the language. Heuristic methods were first proposed before benefiting from advances in pattern mining with sound and complete methods [35]. Indeed, the principle of reducing the search space is very similar to the previous property for constraint-based pattern mining:

Property 2 (Safe pruning). Given a candidate pattern set $S \subseteq \mathcal{L}$ such that $\mathcal{B}est_k(\mathcal{L}, \mathcal{D}, R) \subseteq S$, we have: $\mathcal{B}est_k(\mathcal{L}, \mathcal{D}, R) = \mathcal{B}est_k(S, \mathcal{D}, R)$.

As in the previous section, the goal is to dynamically reduce the candidate pattern set during the search. For this purpose, a branch-and-bound approach can be considered i.e., the best current solution is gradually refined to derive a temporary pruning condition. The progress in the search space exploration improves the current solution that improves the current pruning condition. Table 6 illustrates this principle on the mining of the 3 most frequent itemsets. Once the first solution is computed from the pattern TR, a first pruning condition is derived: $freq(\varphi, \mathcal{D}) < 2$. For example, this pruning condition eliminates the pattern TRDH which has only a frequency of 1. Then the pruning condition is improved when patterns having a higher frequency are added to the current solution. Recent CP frameworks offer more generic solving methods [9,33].

Table 6. Top-3 frequent itemset mining

Pattern	Current top-3	Minimal frequency threshold
$\emptyset$	$\{\emptyset\}$	-
T	$\{\emptyset, \mathrm{T}\}$	-
TR	$\{\emptyset, \mathrm{T}, \mathrm{TR}\}$	2
TRD	$\{\emptyset, \mathrm{T}, \mathrm{TR}, \mathrm{TRD}\}$	2
TD	$\{\emptyset, \mathrm{T}, \mathrm{TD}\}$	3
R	$\{\emptyset, \mathrm{T}, \mathrm{TD}, \mathrm{R}\}$	3
RD	$\{\emptyset, \mathrm{T}, \mathrm{TD}, \mathrm{R}, \mathrm{RD}\}$	3
D	$\{\emptyset, \mathrm{T}, \mathrm{D}\}$	4

5.2 Diversity Issue

Unfortunately, the best patterns for these preferences are sometimes too obvious for the user. In the case of top-k frequent pattern mining, the mined patterns are too general (in particular, the empty set is not interesting). Besides, the best patterns are often very similar to each other and are not representative of the diversity of the language. Instead of using a single criterion, it is possible to combine several preference relations. For instance, given n measures $m_1, \ldots, m_n$, skyline patterns are the most preferred patterns according to the relation $(\varphi R_{m_1,\ldots,m_n} \gamma) \Leftrightarrow (\forall i \in \{1, \ldots, n\})(\varphi R_{m_i} \gamma)$ [36]. Table 7 illustrates this notion with frequency and area. R is not mined even if it has the same frequency than TD because R is dominated by TD which has a higher area. It is easy to see that a best pattern according to one criterion (like TRD for area) does not remain a skyline pattern due to the patterns TD or RD (which have higher frequencies).

Table 7. Skyline patterns for frequency and area

Pattern	Frequency	Area	Pattern	Frequency	Area
∅	**5**	**0**	RD	**3**	**6**
T	**4**	**4**	RH	1	2
R	3	3	DH	1	2
D	**4**	**4**	TRD	2	6
H	2	2	TRH	1	3
TR	2	4	TDH	1	3
TD	**3**	**6**	RDH	1	3
TH	2	4	TRDH	1	4

Early work proposed to reduce the number of resulting patterns by limiting redundancy among those patterns, using so-called condensed representations. More precisely, many works in the literature focus on reducing the number of patterns without loss of information [37,38]. Instead of mining all frequent patterns, their goal is to extract a subset that allows to regenerate all patterns. For example with the maximal frequent patterns with respect to inclusion (i.e., {TH, TRD}), it is easy to regenerate all the patterns whose frequency is at least 2 [4]. Indeed, a pattern having a frequency greater than 2 (say R) is a subset of at least one maximal frequent pattern (here, R $\subseteq$ TRD). However, for regenerating the exact frequency of each pattern, it is necessary to retain more patterns (1 per equivalence class). These patterns (i.e., {∅, T, D, TD, RD, TH, TRD}) are said to be closed [39]. Thereby, the frequency of R can be deduced from that of RD as R $\subseteq$ RD. Note that the notion of closed patterns is strongly linked to that of concept in Formal Concept Analysis [40]. Figure 4 depicts a maximal border separating the frequent patterns from others and plots each equivalence class.

Interestingly, the condensed representations are just a special case of preference-based pattern mining. Maximal frequent patterns stem from the following preference relation: $(\varphi R_{max} \gamma) \Leftrightarrow (\varphi \supset \gamma)$ while closed frequent patterns are obtained with $(\varphi R_{max} \gamma) \Leftrightarrow (\varphi \supset \gamma \wedge \mathit{freq}(\varphi, \mathcal{D}) = \mathit{freq}(\gamma, \mathcal{D}))$. Let us come back to our running example, the set of maximal frequent patterns is $\mathit{Best}(\mathit{Th}(2^{\mathcal{I}}, \mathcal{D}, \mathit{freq}(\varphi, \mathcal{D}) \geq 2), \mathcal{D}, R_{max}) = \{\text{TH}, \text{TRD}\}$.

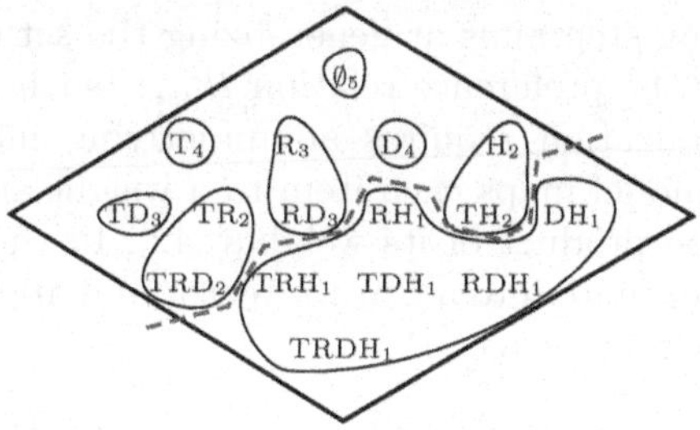

Fig. 4. Equivalence classes of frequency and maximal border (when the minimal frequency threshold is 2)

Limits. Preference-based pattern mining has many advantages. It reduces the number of mined patterns and it focuses on the most preferred patterns. However, although end users no longer have to set difficult thresholds (contrary to constraint-based pattern mining), it remains difficult to explicitly formulate their preferences.

6 Interactive Pattern Mining

In practice, it is difficult for the user to express their interest by stating either a constraint or a preference relation. Several works including [11,41] focus on interactive learning of user preferences. The idea is to submit patterns to the end user and to benefit from their feedback to better target their expectations (see Sect. 6.1). This interactive process requires a short loop with a rapid interaction between the mining system and the user and in particular, it raises the challenge of mining instantly relevant patterns (see Sect. 6.2).

6.1 Learning a User Preference Model from Patterns

Assuming that the user has a preference relation over patterns denoted by R_{user}, interactive pattern mining aims at finding this relation and at the same time, discovering relevant patterns with respect to this relation. Most methods follow a general framework that iterates three steps [11]:

1. **Mine:** The goal of this step is of course to provide relevant patterns to the user. If the first iteration does not rely on the user's interest, the challenge from the second iteration is to integrate the current user preference relation R_{user_i} for extracting high quality patterns.
2. **Interact:** This step captures the view of the user about patterns in the form of *implicit feedback* (e.g., observation time of a pattern or clicks) or *explicit feedback* (e.g., rates or pairwise preferences) where explicit feedback provides more accurate information. Basically, if the user indicates that φ is preferred to γ, $\varphi R_{user} \gamma$ is added to user feedback $\mathcal{F}$. With a rating, if the user gives a better rating for φ than for γ, $\varphi R_{user} \gamma$ is also added to user feedback $\mathcal{F}$.

3. **Learn:** The learning step aims at generalizing the set of user feedback $\mathcal{F}$ to iteratively improve the preference relation R_{user_i} such that $\lim_{i\to\infty} R_{user_i} = R_{user}$. This generalization requires an underlying model. For instance, a weighted product model maps each item to a weight and considers the score of an itemset as the product of its weights [42]. In the same way, a feature space model maps each item to a feature vector and applies a learning to rank approach on this feature space [41,43].

One of the main challenges of this cycle is its *active learning* nature [41]. Indeed, the improvement of the preference model requires an adequate choice of patterns that are provided to the user. If the mining step still provides similar patterns, the preference model cannot be improved. It means that the mining step has to select varied and representative patterns (in addition to extract patterns having a high quality according to R_{user_i}).

Another challenge is the choice of the preference model which determines the user representation [42,43]. This model requires to include a large set of features not to miss the one that will capture the user interest. But, if this model is too complex, it is really difficult to integrate it into the mining step.

6.2 Pattern Sampling

Interactive nature imposes a constraint of few seconds to extract patterns. Sound and complete mining methods cannot obtain the relevant patterns in such a short response time. Heuristic methods focus on the approximate search of the best patterns with respect to a preference relation. Therefore, they often focus on the same part of the pattern language $\mathcal{L}$ which is often suboptimal and which contains slightly different patterns. However, as explained in the previous section, diversity is a crucial point for interactive methods. It is important to present to the user a set of varied patterns at each step to improve their vision of data and to help the system to learn their interestingness from new feedback.

Pattern sampling is a new mining method that guarantees a very fast extraction and a high diversity between mined patterns. It aims at accessing the pattern space $\mathcal{L}$ by an efficient sampling procedure simulating a distribution $\pi : \mathcal{L} \rightarrow [0,1]$ that is defined with respect to some interestingness measure m: $\pi(.) = m(.)/Z$ where Z is a normalizing constant. In this way, the user has a fast and direct access to the entire pattern language and with no parameter (except possibly the sample size). As constraint-based and preference-based pattern mining, pattern sampling problem has been declined for different languages like itemsets [12] and graphs [44], and different interestingness measures [12,44] including support, area, discriminative measures and utility measures.

Problem 3 (Pattern sampling). Given a language $\mathcal{L}$, a dataset $\mathcal{D}$ and an interestingness measure m, pattern sampling aims at randomly picking k patterns from $\mathcal{L}$ according to a distribution proportional to m in the dataset $\mathcal{D}$:

$$\mathit{Samp}_k(\mathcal{L}, \mathcal{D}, m) = \{\varphi_1, \ldots, \varphi_k \sim m(\mathcal{L}, \mathcal{D})\}$$

The philosophy of the operator $\mathcal{S}amp$ is very different from those seen previously (i.e., $\mathcal{T}h$ and $\mathcal{B}est$). First, the operator applied several times on the same operands does not necessarily return the same patterns. Second, any pattern of $\mathcal{L}$ can be returned as soon as its value is greater than 0 for the considered measure. Considering the toy example of Table 1 with $\mathcal{S}amp(2^{\mathcal{I}}, \mathcal{D}, supp)$, as the frequency of T is 4 and that of TH is only 2, T has twice more chance to be picked than TH.

Pattern sampling was first investigated on graphs [44] and later on itemsets [12]. Usually, new pattern mining techniques are not introduced on graphs whose structure is complex and leads to a very large pattern language. However, the complexity of sampling techniques does not depend on the size of the language and then, pattern sampling is a natural response to large languages [45]. There are two main families of pattern sampling techniques. Markov Chain Monte Carlo (MCMC) method [44] uses a random walk on the partially ordered graph formed by the pattern language. With such a stochastic simulation, it is difficult to set the equilibrium distribution with the desired properties and the convergence to the stationary distribution within an acceptable error can be slow. Two-step random procedure [12] samples patterns exactly and directly without simulating stochastic processes. Basically, this procedure randomly selects a transaction according to a first distribution and then, it selects a pattern from this transaction according to a second distribution. Clearly, the choice of these two distributions allows a fine control of the produced patterns in order to consider different interestingness measures (e.g., area or contrast measure). This method is particularly effective for drawing patterns according to support or area (linear with the size of the dataset). But it turns out quadratic or worse for some measures (like contrast measures) requiring the drawing of several transactions in the first step. In addition to its good time complexity, pattern sampling has good properties for building more sophisticated pattern-based models [46,47].

7 Conclusion

This paper provides a very short and partial overview of pattern mining and it should be completed by discussions about practical use cases and evaluation (which remains a critical issue as for all unsupervised discovery methods). Nevertheless, pattern mining clearly follows trends that can be roughly summarized:

- **Faster:** The first concern of pattern mining was to develop algorithms to quickly return responses despite a huge search space. The speed of execution justified to extract *frequent* patterns even if they have limited interest for end users. Most work is still based on how to optimize algorithms for effectively extract the desired patterns. Recently, the arrival of interactive pattern mining renewed the interest of a short time response (but completeness is no longer required).

- **Better:** The passage of frequent pattern mining to constraint-based pattern mining was a very important first step to improve the quality of mined patterns. Preference-based pattern mining goes a little further by focusing on the patterns maximizing a quality criterion. All these methods are clearly intended to benefit from explicit knowledge provided by the user. Interactive pattern mining takes an opposite view by directly learning the user's interest from their feedback.
- **Easier:** The input parameters of mining methods perfectly illustrate the movement of simplification. The first users were asked to select the appropriate algorithm for each type of dataset. After, the user just had to formulate their constraints and thresholds. Finally, preference-based pattern mining withdrew the thresholds. Currently, interactive pattern mining even removes the need for the user to explicitly specify their interest. Meanwhile, this simplification in the problem specification was accompanied by works on the simplification of solving methods thanks to generic solvers.

We think that the direction of pattern mining is the same as that followed by related fields of Computer Science (e.g., Databases or Information Retrieval). Pattern mining is moving towards exploratory data analysis where new search methods are less data-centric and more user-centric.

Acknowledgments. The author would like to thank Bruno Crémilleux, Arnaud Giacometti and Marc Plantevit for many fruitful discussions. The author would also like to thank the anonymous reviewers and Patrick Marcel for their helpful comments that greatly contributed to improve the final version of the paper.

References

1. Agrawal, R., Imieliński, T., Swami, A.: Mining association rules between sets of items in large databases. ACM Sigmod Record **22**(2), 207–216 (1993)
2. Giacometti, A., Li, D.H., Marcel, P., Soulet, A.: 20 years of pattern mining: a bibliometric survey. ACM SIGKDD Explor. Newslett. **15**(1), 41–50 (2014)
3. Mitchell, T.M.: Generalization as search. Artif. Intell. **18**(2), 203–226 (1982)
4. Mannila, H., Toivonen, H.: Levelwise search and borders of theories in knowledge discovery. Data Min. Knowl. Disc. **1**(3), 241–258 (1997)
5. Nijssen, S., Zimmermann, A.: Constraint-Based Pattern Mining, pp. 147–163. Springer, Cham (2014)
6. Bonchi, F., Lucchese, C.: Extending the state-of-the-art of constraint-based pattern discovery. Data Knowl. Eng. **60**(2), 377–399 (2007)
7. Pei, J., Han, J.: Constrained frequent pattern mining: a pattern-growth view. ACM SIGKDD Explor. Newslett. **4**(1), 31–39 (2002)
8. Soulet, A., Crémilleux, B.: Mining constraint-based patterns using automatic relaxation. Intell. Data Anal. **13**(1), 109–133 (2009)
9. Negrevergne, B., Dries, A., Guns, T., Nijssen, S.: Dominance programming for itemset mining. In: 2013 IEEE 13th International Conference on Data Mining, pp. 557–566. IEEE (2013)

10. Ugarte Rojas, W., Boizumault, P., Loudni, S., Crémilleux, B., Lepailleur, A.: Mining (Soft-) skypatterns using dynamic CSP. In: Simonis, H. (ed.) CPAIOR 2014. LNCS, vol. 8451, pp. 71–87. Springer, Cham (2014). doi:10.1007/978-3-319-07046-9_6
11. Leeuwen, M.: Interactive data exploration using pattern mining. In: Holzinger, A., Jurisica, I. (eds.) Interactive Knowledge Discovery and Data Mining in Biomedical Informatics. LNCS, vol. 8401, pp. 169–182. Springer, Heidelberg (2014). doi:10.1007/978-3-662-43968-5_9
12. Boley, M., Lucchese, C., Paurat, D., Gärtner, T.: Direct local pattern sampling by efficient two-step random procedures. In: Proceedings of the 17th ACM SIGKDD International Conference on Knowledge Discovery and Data Mining, pp. 582–590. ACM (2011)
13. De Raedt, L., Zimmermann, A.: Constraint-based pattern set mining. In: SDM, SIAM, pp. 237–248 (2007)
14. Guns, T., Nijssen, S., De Raedt, L.: Itemset mining: a constraint programming perspective. Artif. Intell. **175**(12), 1951–1983 (2011)
15. Srikant, R., Agrawal, R.: Mining quantitative association rules in large relational tables. In: Acm Sigmod Record. vol. 25, pp. 1–12. ACM (1996)
16. Kaytoue, M., Kuznetsov, S.O., Napoli, A.: Revisiting numerical pattern mining with formal concept analysis. In: International Conference on Artificial Intelligence (IJCAI 2011) (2011)
17. Agrawal, R., Srikant, R.: Mining sequential patterns. In: Proceedings of the Eleventh International Conference on Data Engineering, pp. 3–14. IEEE (1995)
18. Zhao, Q., Bhowmick, S.S.: Sequential pattern mining: a survey. ITechnical Report CAIS Nayang Technological University Singapore, pp. 1–26 (2003)
19. Jiang, C., Coenen, F., Zito, M.: A survey of frequent subgraph mining algorithms. Knowl. Eng. Rev. **28**(01), 75–105 (2013)
20. Arimura, H., Uno, T.: Polynomial-delay and polynomial-space algorithms for mining closed sequences, graphs, and pictures in accessible set systems. In: SDM (2009)
21. Agrawal, R., Srikant, R., et al.: Fast algorithms for mining association rules. In: Proceeding of 20th International Conference Very Large Data Bases, VLDB, vol. 1215, pp. 487–499 (1994)
22. Geng, L., Hamilton, H.J.: Interestingness measures for data mining: a survey. ACM Comput. Surv. (CSUR) **38**(3), 9 (2006)
23. Vreeken, J., Tatti, N.: Interesting Patterns. In: Aggarwal, C.C., Han, J. (eds.) Frequent Pattern Mining, pp. 105–134. Springer, Cham (2014). doi:10.1007/978-3-319-07821-2_5
24. Leman, D., Feelders, A., Knobbe, A.: Exceptional model mining. In: Daelemans, W., Goethals, B., Morik, K. (eds.) ECML PKDD 2008. LNCS, vol. 5212, pp. 1–16. Springer, Heidelberg (2008). doi:10.1007/978-3-540-87481-2_1
25. Webb, G.I.: Self-sufficient itemsets: an approach to screening potentially interesting associations between items. ACM Trans. Knowl. Discov. Data (TKDD) **4**(1), 3 (2010)
26. Tan, P.N., Kumar, V., Srivastava, J.: Selecting the right interestingness measure for association patterns. In: Proceedings of the Eighth ACM SIGKDD International Conference on Knowledge Discovery and Data Mining, pp. 32–41 (2002)
27. Winslett, M.: Interview with Rakesh Agrawal. SIGMOD Rec. **32**(3), 83–90 (2003)
28. Fayyad, U.M., Piatetsky-Shapiro, G., Uthurusamy, R.: Summary from the KDD-03 panel: data mining: the next 10 years. ACM SIGKDD Explor. Newslett. **5**(2), 191–196 (2003)

29. Han, J., Cheng, H., Xin, D., Yan, X.: Frequent pattern mining: current status and future directions. Data Min. Knowl. Disc. **15**(1), 55–86 (2007)
30. Zaki, M.J., Parthasarathy, S., Ogihara, M., Li, W., et al.: New algorithms for fast discovery of association rules. KDD **97**, 283–286 (1997)
31. Goethals, B.: Survey on Frequent Pattern Mining. University of Helsinki (2003)
32. Cerf, L., Besson, J., Robardet, C., Boulicaut, J.F.: Data peeler: contraint-based closed pattern mining in n-ary relations. In: SDM, vol. 8, pp. 37–48. SIAM (2008)
33. Ugarte, W., Boizumault, P., Loudni, S., Crémilleux, B.: Modeling and mining optimal patterns using dynamic CSP. In: 27th International Conference on Tools with Artificial Intelligence (ICTAI), pp. 33–40. IEEE (2015)
34. Fu, A., Kwong, R., Tang, J.: Mining N-most interesting itemsets. In: Raś, Z.W., Ohsuga, S. (eds.) ISMIS 2000. LNCS, vol. 1932, pp. 59–67. Springer, Heidelberg (2000). doi:10.1007/3-540-39963-1_7
35. Herrera, F., Carmona, C.J., González, P., Del Jesus, M.J.: An overview on subgroup discovery: foundations and applications. Knowl. Inf. Syst. **29**(3), 495–525 (2011)
36. Soulet, A., Raïssi, C., Plantevit, M., Cremilleux, B.: Mining dominant patterns in the sky. In: 2011 IEEE 11th International Conference on Data Mining, pp. 655–664. IEEE (2011)
37. Calders, T., Rigotti, C., Boulicaut, J.-F.: A survey on condensed representations for frequent sets. In: Boulicaut, J.-F., Raedt, L., Mannila, H. (eds.) Constraint-Based Mining and Inductive Databases. LNCS, vol. 3848, pp. 64–80. Springer, Heidelberg (2006). doi:10.1007/11615576_4
38. Hamrouni, T.: Key roles of closed sets and minimal generators in concise representations of frequent patterns. Intell. Data Anal. **16**(4), 581–631 (2012)
39. Pasquier, N., Bastide, Y., Taouil, R., Lakhal, L.: Discovering frequent closed itemsets for association rules. In: Beeri, C., Buneman, P. (eds.) ICDT 1999. LNCS, vol. 1540, pp. 398–416. Springer, Heidelberg (1999). doi:10.1007/3-540-49257-7_25
40. Ganter, B., Wille, R.: Formal Concept Analysis: Mathematical Foundations. Springer Science & Business Media, Heidelberg (2012)
41. Dzyuba, V., Leeuwen, M.V., Nijssen, S., De Raedt, L.: Interactive learning of pattern rankings. Int. J. Artif. Intell. Tools **23**(06), 1460026 (2014)
42. Bhuiyan, M., Mukhopadhyay, S., Hasan, M.A.: Interactive pattern mining on hidden data: a sampling-based solution. In: Proceedings of the 21st ACM International Conference on Information and Knowledge Management, pp. 95–104. ACM (2012)
43. Rueping, S.: Ranking interesting subgroups. In: Proceedings of the 26th Annual International Conference on Machine Learning, pp. 913–920. ACM (2009)
44. Hasan, M.A., Zaki, M.J.: Output space sampling for graph patterns. PVLDB **2**(1), 730–741 (2009)
45. Bendimerad, A.A., Plantevit, M., Robardet, C.: Unsupervised exceptional attributed sub-graph mining in urban data. In: IEEE 16th International Conference on Data Mining (ICDM), pp. 21–30. IEEE (2016)
46. Moens, S., Boley, M., Goethals, B.: Providing concise database covers instantly by recursive tile sampling. In: Džeroski, S., Panov, P., Kocev, D., Todorovski, L. (eds.) DS 2014. LNCS, vol. 8777, pp. 216–227. Springer, Cham (2014). doi:10.1007/978-3-319-11812-3_19
47. Giacometti, A., Soulet, A.: Frequent pattern outlier detection without exhaustive mining. In: Bailey, J., Khan, L., Washio, T., Dobbie, G., Huang, J.Z., Wang, R. (eds.) PAKDD 2016. LNCS, vol. 9652, pp. 196–207. Springer, Cham (2016). doi:10.1007/978-3-319-31750-2_16

Scalability and Realtime on Big Data, MapReduce, NoSQL and Spark

Pedro Furtado(✉)

Departamento de Engenharia Informatica and Centro de Informatica e Sistemas da Universidade de Coimbra, Universidade de Coimbra, Polo II, 3030-290 Coimbra, Portugal
pnf@dei.uc.pt

Abstract. Big data platforms strive to achieve scalability and realtime for query processing and complex analytics over "big" and/or "fast" data. In this context, big data warehouses are huge repositories of data to be used in analytics and machine learning. This work discusses models, concepts and approaches to reach scalability and realtime in big data processing and big data warehouses. The main concepts of NoSQL, Parallel Data Management Systems (PDBMS), MapReduce and Spark are reviewed in the context of scalability. The first two offering data management, the last two adding flexible and scalable processing capacities. We also turn our attention to realtime data processing, lambda architecture and its relation with scalability, and we revisit our own recent research on the issue. Three approaches are included that are directly related to realtime and scalability: the use of a realtime component in a data warehouse, parallelized de-normalization for scalability and execution tree sharing for scaling to simultaneous sessions. With these models and technologies we revisit some of the major current solutions for data management and data processing with scalability and realtime capacities.

Keywords: Big data · Realtime · Scalability · NoSQL · MapReduce · DBMS · Hadoop · Spark · Spark-SQL

1 Introduction

Big data is commonly characterized as data that exhibits some or all of the "V" properties. This includes Volume (the quantity of data), Variety (the type of data), Velocity (the speed at which the data is generated and needs to be processed), Variability (varying structure of the data), and Veracity (quality of captured data). A narrower definition of big data is data whose sizes are beyond the ability of commonly used software tools to capture, curate, manage, and process them within a tolerable elapsed time [39]. Others stress the fact that the value of big data is what can be learned from it. Big data analytics can reveal knowledge previously hidden inside enormous amounts of data. This knowledge is now within reach. Since the first massive scalability software platforms appeared, it became possible for anyone to deploy a massive data processing platform, analyze data of huge sizes and discover new knowledge from that data.

P. Marcel and E. Zimányi (Eds.): eBISS 2016, LNBIP 280, pp. 79–104, 2017.
DOI: 10.1007/978-3-319-61164-8_4

Big data also appears related to runtimes constraints. Data may exceed the capacity of the systems that were planned to process it to answer requests within a useful time frame or deadline. This definition of big data emphasizes a deadline, which is totally dependent on the context and objectives of the system and its users.

A big data warehouse is a massive amount of historical data used for analysis. Based on this definition, many big data challenges can be considered big data analytics over a big data warehouse, although possibly not a conventional one. We see big data warehouses as a kind of big data for analytics with more or less of the "V" properties, but where the main challenge is to provide processing time guarantees regardless of the size of the dataset. One objective is to design systems that are able to process huge amounts of data (scalability) within bounded response time limits (realtime) (to handle Volume and Velocity efficiently). In this work we discuss architectures, models and technologies that enable big data analytics and fulfil scalability and realtime objectives. In this context, we review concepts such as NoSQL and MapReduce, and technologies such as Hadoop and Spark. Although the remaining Vs (Variety, Variability, Veracity) are not the central concern in this work, they should also be present and should be part of the big data analytics solutions.

The first concern of data warehousing and data analysis solutions for big data should be that of storing and managing huge amounts of data with scalability. This need motivates our initial review of parallel database management systems (PDBMS), NoSQL and Hadoop, Hadoop being a modern massive scalability platform. Besides storage and basic data management, data processing for massive data is another requirement for handling big data. This is why we review the concept of MapReduce and Spark, a technology using MapReduce concepts. Finally, if we have enormous amounts of data and/or an enormous data ingest rate from which we need to refresh some model frequently, there is the concern that processing should also be realtime. In this context we review the lambda architecture and approaches to achieve realtime and scalability in the presence of high data ingestion and massive numbers of sessions accessing the data.

In terms of use cases, big data challenges and big data warehouses appear in various contexts. Examples related to science include The Sloan Digital Sky Survey (SDSS), which collects astronomical data at a rate of about 200 GB per night, and its successor, the Large Synoptic Survey Telescope is expected to multiply that rate many times [41]. NASA's Center for Climate Simulation (NCCS) stores 32 petabytes of climate observations and simulations on the Discover supercomputing cluster [42]. Google's DNAStack compiles and organizes DNA samples of genetic data from around the world to identify diseases and other medical defects [43].

In retail, Walmart's databases contain an estimate of 2.5 petabytes of data and handle more than 1 million customer transactions every hour [41]. In contexts such as manufacturing, cyber-physical systems and m-health, sensors are able to collect huge amounts of information at high rate, and huge numbers of "things" need to exchange, share and manage data in the internet-of-things cloud. It is not only a case of volume and velocity, the fraction of inaccurate data also increases as the volume of data increases, therefore intelligent tools need to control the accuracy and veracity of the sensed data and information discovered from there [44]. Open research issues about big data on cloud computing are also discussed in [40].

In this chapter we discuss data management scalability architectures and models in the context of data analytics, and including recent developments. After revisiting parallel database management systems and concepts such as NoSQL and MapReduce, we review the concepts that work in scalable manner to enable data processing scalability in Spark (RDDs, Spark-SQL, datasets and data frames), and some of their properties related to efficiency. After that we review the concept of realtime analytics and the lambda architecture that is most related to that concept. Finally, we briefly review our own research efforts on approaches to add realtime, volume and session scalability to big data warehouses.

The chapter is organized as follows. Section 2 discusses big data architectures, scalability and MapReduce. Section 3 discusses Spark and Spark-SQL, and Sect. 4 discusses realtime scalable big data analytics. Section 5 concludes the chapter.

2 Big Data Architectures and Scalability

The first major concern when dealing with big data is how to store and manage it efficiently. One should be able to store any amount of data and the data management architecture should be able to scale to many nodes automatically when needed. It should also be able to return data quickly, in spite of the amount of data. Given the importance of scalability for handling big data efficiently, and also the relevance of variability and variety in the context of big data, a set of architectures emerged with characteristics designed to achieve those objectives (Volume, Variety, Variability and Velocity). Some of the most influential models include NoSQL and key-value data stores, parallel database management systems and MapReduce.

2.1 NoSQL and Key-Value Data Stores

The term NoSQL derives from "Not Only SQL". It encompasses a large and diverse set of architectures and technologies whose objective is to provide massive scalability and performance to handle big data. Another important characteristic is support of data structure flexibility, variety and variability. Those are data processing issues that relational databases weren't designed to address specifically. NoSQL usefulness also comes from the realization that many new data sets are not all alike. For instance, graph databases are well suited for those situations where data is organized by relationships instead of by row or document.

Many NoSQL systems do not support some of the relational operations existing in relational databases (e.g. joins, or representation as relations), and many trade consistency for scalability and availability [45]. Not supporting transactions natively and not guaranteeing the full set of ACID properties (atomicity, consistency, isolation and durability) saves a lot of housekeeping overhead and provides improved performance and flexibility, although it may prevent the system from being used in consistency-critical contexts and applications. However, the universe of contexts and applications where big data management systems can be used is very wide, and perhaps most of them are not consistency-critical.

NoSQL includes various (sub-)architectures (e.g. key-value caches, key-value stores, document stores). Next we present an abstract model fitting some NoSQL data stores, and in particular Apache Cassandra™ [37, 57]. Our abstract model manages data organized as sets of tuples structured as key-value pairs, where the key is the identifier used for accessing the data, and the value can be very different things depending on the particular NoSQL data store, e.g. a string, some tuple, a json document, a blob. The value part of the tuple may or may not be typed, also depending on the particular NoSQL data store. The tuple may be a set of (attribute:value) pairs. There may be a schema, but tuples can vary in the actual values that they have. Data can also be organized physically in different ways (e.g. as raw data, json files or column-wise compressed storage).

Figure 1 illustrates the scalable NoSQL data store model with 4 nodes. Nodes can be added or removed as necessary, and the system re-organizes itself to load the data and processing into the number of available nodes. From the point of view of data management over multiple nodes, this NoSQL data store model is like a massively parallel hash-table, indexing the key into nodes based on hash intervals (another level of indexing by key will find specific tuples faster within the target node). Each node becomes responsible for an interval or set of hash values. The NoSQL engine offers the following main access primitives: PUT, GET, RANGE, FILTER AND SCAN. The primitive *put (key, value)* hashes the key and determines the node whose hash interval contains that key value, to store the value. The primitive *value = get(key)* hashes the key, finds the node with that hash value and retrieves the tuple. SCAN scans all tuples, and RANGE searches in a range of key values. FILTER is like a scan but filters rows based on some condition.

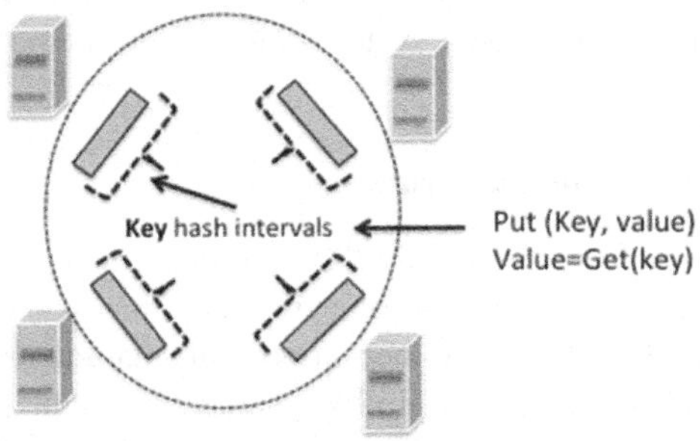

Fig. 1. Abstraction of a scalable NoSQL data store

Most NoSQL data stores do not support joins and other complex relational operations. This relieves NoSQL from heavy time-consuming operations, but either the application does not need to use such operations, or there must be some replacement. One important difference in some NoSQL systems is the support for nested datasets (e.g. questions/answers in a forum) that reduces the need to do joins. However, there are still many cases in practical applications and contexts where joins or group-by aggregations are required. In those cases the operations need to be done in the client or by integrating the NoSQL data store with a massive scalability platform such as Hadoop and Spark, to process those parts. This is in contrast to a RDBMS that processes and optimizes joins and aggregations by itself.

Apache Cassandra™ is an example of a NoSQL data store [57] with a model similar to the one just described. Cassandra is completely decentralized, fault tolerant and any individual node can fail without halting the system. It automatically replicates data into a number of nodes to avoid loosing data or availability if a node fails. Nodes can be added or removed easily, with the data and hash intervals being re-arranged automatically to handle the new node(s). With this organization, NoSQL data stores are quite useful in applications such as social networks with massive amounts of users that retrieve and edit data based on a key.

Bigtable™ is another NoSQL data store, one that works on top of Google File System™ and which has a different organization from the previously model. Google™'s reasons for developing Bigtable include scalability and better control of performance. It was developed around 2004 and is used by several Google™ applications, such as web indexing and MapReduce, Google Maps, YouTube, Gmail and Google Cloud Platform. A Bigtable is a sparse, distributed, persistent multidimensional sorted map indexed by a row key, column key, and a timestamp, and where each value in the map is an un-interpreted array of bytes ((row:string, column:string, time:int64) → string) [46]. The data is maintained in lexicographic order by row key, and a table is dynamically partitioned into row ranges called tablets, which are also the unit for distribution, scaling and load balancing. The row key or name is a string indexing the row, which is made of a set of column families. A column family is a set of column keys, and a column key is a pair family:qualifier. Column family names must be printable, but qualifiers may be arbitrary strings. The Bigtable API includes functions for creating and deleting tables and column families. A Bigtable cluster stores a number of tables. Each table consists of a set of tablets, and each tablet contains all data associated with a row range. Initially, each table consists of just one tablet. As a table grows, it is automatically split into multiple tablets, each approximately 100–200 MB in size by default [46]. Tablet location information is kept in a B^+-tree-like structure. Tablet distribution and load balancing over a dynamically adjustable number of nodes is also managed by Bigtable software. Bigtable is designed to scale across "hundreds or thousands of machines, and to make it easy to add more machines and automatically start taking advantage of those resources without any reconfiguration" [47].

There are many other NoSQL models and systems that we do not discuss in this work. For instance, graph databases [48, 49] are effective to represent data that is mostly relationships between entities. For information on other types of data stores and NoSQL models, please refer to survey, such as [50–52].

NoSQL data stores were developed to handle storage and access to huge amounts of data efficiently. Most importantly, they support data structure flexibility, variety and variability, a major departure from relational databases that require data to fit a fixed pre-defined schema and need to load, binarize and compress data. NoSQL architectures are more flexible by not requiring fixed schemas and fitting varied data structures. But scalable architectures evolved initially within the context of relational database models, and many of the concepts used in modern scalability platforms evolved from those of parallel database management systems. That is why we review parallel database management systems next.

2.2 Parallel Database Management Systems

Parallel database management systems evolved in the eighties and nineties from database management systems as a way to give scalability to data processing in relational databases. The typical reasons for using parallel systems to process data are to achieve high performance, high availability and scaling. Performance benefits are manifested in execution times and balancing of the load among existing processors. Availability, on the other hand, is manifested in tolerance to faults in any component, redundancy and consistency. A speedup of n in a parallel system means that n times as much hardware will get a reduction of execution time by a factor of n, while a scale-up of n means the hardware will allow you to process n times as much data in the same elapsed time. There are four main factors limiting speedup and scale-out in a parallel query processing system: The start-up time, which is the time each processor needs to start processing data, interference between processors, since they are executing data simultaneously and sharing resources, increased communication costs, since data has to be exchanged at certain points, and data skew, which has to do with unbalance when the data and processing are divided between processors for faster operation.

Data partitioning is the process of distributing the tuples of a relation over several disks or nodes or processors, to allow the parallel query processor to exploit data processing parallelism. The most common data partitioning algorithms are round-robin, range-based or hash-based. Data skew most frequently results from the distribution of the data during data partitioning. For instance, if one node/processor is left with most of the data, then the benefits of parallelization decrease due to unbalanced load. Skew can be minimized using skew-resistant hashing and/or having many fixed sized data chunks allocated in a balanced way.

Parallel query processing is based on execution of a parallel execution plan. The parallel execution plan itself results from transformation of a relational query plan into a parallel execution plan. The parallel execution plan should minimise a cost function that captures resource consumption costs. Parallel execution control strategies are classified as either control flow-based or dataflow-based. In control flow-based approaches, a single control node controls all processes related to a query, scheduling, starting and synchronising all processes. In a dataflow strategy there is no centralized execution control, as data flowing between processes triggers its processing (data driven execution control). In dataflow architectures, parallelization of relational operators can be achieved by means of parallelized data streams flowing in and out of existing sequential relational operators available in all processors. Each relational operator has input ports on which input tuples arrive and output ports to place the output stream of tuples. Given this organization, parallel dataflow works by partitioning and merging data streams from and into those ports. Further references on this issue include Volcano [53], a parallel query evaluation system based on dataflow control, surveys on distributed query processing approaches [54] and on adaptive query processing [55].

A concept that is also useful to understand parallel query execution is "additivity" or "composability". Most relational operators are highly parallelizable and composable or additive. Being composable or additive means that the data to be processed can be divided into parts, executed in parallel and then merged. For instance, A + B + C + D can be executed independently as AB = (A + B) and CD = (C + D) in different processors, and the result can be merged as AB + CD. In intra-query parallelism each node processes operators on its share of the data and the result is merged afterwards. Horizontal intra-query parallelism refers to dividing the data into multiple nodes, so that each node will process part of the data. A node can be a physical machine, a processor, a core or a thread, for instance. The data is partitioned into chunks and may be shipped physically into the nodes following some partitioning strategy, or otherwise it can rest in common storage that nodes access as needed. Each node runs the query or operator over its share of the data, and at the end another query or operator merges the results coming from all nodes. Figure 2 shows a simple example. The original query "sum(a) over Data group by C, D" is ran on each node and another query merges the results coming from the nodes. Most relational operators are highly parallelizable this way, but a few operators, such as join and percentile, require further transformations to be executed in parallel. Please refer to [17, 19, 20] for more details on this subject.

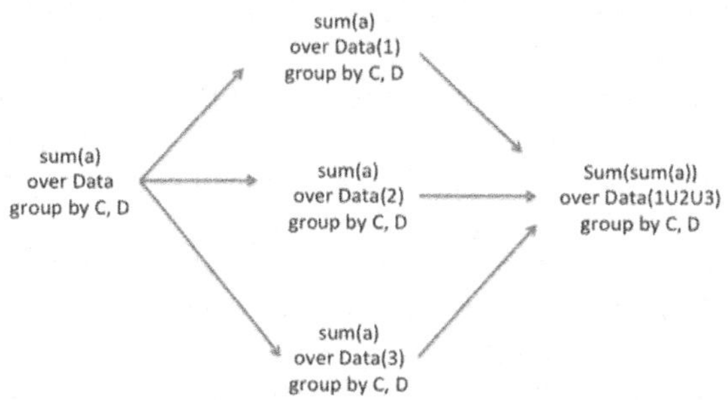

Fig. 2. Illustrating parallel processing of simple query

Fault tolerance and load balancing are two fundamental issues in parallel database systems. In order to optimize load balancing, and therefore performance, you need to divide the dataset into chunks of fixed size (e.g. 100 MB or 1 GB) and use those chunks as a unit for partitioning and replication. Every chunk needs to be replicated into more than one node, to guarantee that if a node fails the other nodes keep a copy and can replace the original node processing the chunk. Automatic load balancing and fault tolerance are guaranteed by this design and the use of an on-demand query processing pattern [20], where the master assigns work over chunks to nodes when they are available. This on-demand parallel execution pattern was applied in parallel database systems, and is also applied more recently in scalability platforms, including Bigtable and Hadoop.

2.3 Joins, Aggregations and Shuffling

Joins can become heavy operations in scalable parallel relational data processing systems, because they need to determine matches of all tuples of one relation with all tuples of the other relation. Consider a relation A(a1, a2) with tuples (1, 'a1'), (2, 'a2'),

(3, 'a3') and a relation B(b1, b2) with tuples (1, 'b1'), (3, 'b2'). The query "select a2, b2 from A join B on (a1 = b1)" should return tuples ('a1' ,'b1') and ('a3' ,'b2'). In a parallel system with two nodes and round-robin placement of the tuples of each relation into the two nodes, the nodes may end up as shown in Fig. 3.

```
Node 1: A: (1,'a1'), (3,'a3')      B: (1,'b1')

Node 2: A: (2,'a2')                B: (3,'b2')
```

Fig. 3. Data placement and join processing

If the join would be executed in parallel without any data exchange between nodes, the result would be only tuple ('a1', 'b1'). This problem is avoided by shipping data tuples between nodes when needed (shuffling), but this shuffling can harm performance significantly if the amount of data to ship is too large and if exchanging data between nodes is costly.

When we can place the data from relations A and B so that the same join key values end up in the same node (equi-partitioning), we will not need to shuffle tuples to process the join. This can be achieved by hashing and assigning equal hash values or hash intervals from both relations to the same node). If we can have balanced equi-partitioning, a dataset of size S divided by *n* nodes will occupy S/n on each node and processing of the join will occur in parallel, expected to take around 1/*n* of the time taken to execute in a single node. But besides the lack of placement flexibility that equi-partitioning requires, many times it is not possible to guarantee that all relations will be equi-partitioned. In those cases repartitioning is needed. Repartitioning is the process of shuffling data on-the-fly so that it becomes equi-placed to proceed with join processing in parallel. Repartitioning can be an expensive operation. However, whenever possible, operations such as early selection and projection are applied prior to repartitioning, to reduce the dataset size that needs to be repartitioned. Another alternative is a broadcast join. In a broadcast join, one of the datasets needs to be broadcasted into all nodes (the smallest dataset). For more details on parallel database management systems, join processing and query processing over parallel relational database management systems, please refer to [14–19].

The concepts that we have just reviewed concerning parallel database management systems were important in the context of relational data but also today in the context of big data with flexible schemas and data variability. In the next sub-section we review how modern scalability platforms work, by introducing Hadoop and MapReduce.

2.4 Hadoop and MapReduce

Modern platforms for processing massive amounts of data, such as Apache's Hadoop™, rely on a user-transparent [Hadoop] Distributed File System (HDFS), with fault tolerance automatically enforced by the file system manager itself. HDFS is a java-based distributed file system that provides scalable and reliable data storage designed to work on a possibly large cluster of commodity nodes. It organizes files as chunks of data with pre-defined data size, then distributes those chunks automatically as the file is copied or written into the HDFS, and does all the house-keeping necessary to keep track of the locations of chunks and to manage them. Chunks are also replicated automatically for fault tolerance.

While the HDFS is a scalable and fault tolerant distributed file system, it is also necessary to have some data processing language and framework to work on partitioned datasets in parallel. MapReduce [56] offers automatic parallelization primitives to be used by programmers in their algorithms, and Hadoop™ offers the support for executing the MapReduce primitives, running parallelized operations on the chunk-partitioned data.

MapReduce is explained for instance in [31]. The input to MapReduce is data from either a flat file, a database, a NoSQL data store or even streaming data. In MapReduce, computation is based on any number of map and/or reduce stages. Map works as follows: map worker threads read the input data and process it by doing row-wise computations, such as filtering the rows based on some conditions. Assigning chunks of data into different workers and having any number of map threads over any number of nodes achieves parallelism and scalability. The Map jobs emit as output (key, value) pairs. Each value of the key in each node will correspond to a specific output file (or buffer). All output datasets with the same key value from all nodes are then merged (concatenated) for further processing. This is the shuffling step, involving possibly-heavy data exchange between nodes. Reducers are workers that will process the shuffled data chunk-wise to merge and aggregate it. The details of the map and reduce operations are programmed by the user.

A simple example of MapReduce is given in Fig. 4, by considering how the query "select sum(x),y from table where z = 2 group by y" would be written in MapReduce.

At this point it is useful to note that the same difficulties faced by parallel database systems when dealing with joins or aggregations or other operations requiring heavy shuffling are also faced by these massive scalability platforms. When processing typical relational data, platforms such as Hadoop™ are in essence working like a parallel database management system, using a user-transparent distributed file system to hold the data and requiring most of the query processing optimizations that traditional parallel database management systems also require.

While Hadoop™ implements a distributed file system (HDFS) and MapReduce, it has some performance limitations, due to the fact that it flushes intermediate data to disk while shuffling, starts next stages of processing after all the previous ones ended, and it uses Java JVM heap space and object serialization for shuffling steps. These limitations have been the focus of much improvement recently. In particular, Apache Spark™ [32] is a framework that applies the MapReduce paradigm on top of file systems or cluster managers such as Hadoop™, but at the same time it tries to optimize the use of memory and the processing of data, minimizing the use of disk storage when memory is available, improving serialization, offering off-JVM heap memory and also efficient file formats (e.g. Parquet files [34]). The next section presents Spark and Spark-SQL, an efficient SQL interface on top of Spark.

```
Map: input=table(x,y,z)
for each line or table(x,y,z)
          if(z==2)
               produce pair (key=y, value=x)
               emit(y, x)
Shuffle: input=set of (y,x)
Send all values with same y (key) value into to a
reducer group
Reduce: input= set of (y,x)
          sumX=0;
          for each pair(y, x)
                sumX+=x;
          return sumX,y
```

Fig. 4. MapReduce for query (select sum(x),y where z = 2 group by y)

3 Spark and Spark-SQL

Spark [33] is a framework for running applications on a parallel/distributed setup with parallelization and scalability. You can use either Scala, Python or Java to program Spark™ and it includes the necessary primitives to run MapReduce tasks. The basis for most processing in Spark™ is the "Resilient Distributed Dataset" (RDD), a distributed structure holding data, preferably in-memory. Spark is easy to use. The next python example illustrates this. It reads a dataset organized with comma-separated fields, computes and prints statistics for each column. In this code fragment, dataset, dataArray and statistics are all RDDs and can be scaled to any number of nodes if necessary (Fig. 5).

The Resilient Distributed Dataset (RDD) is described in [30] as an immutable node-partitioned set of data that is operated in parallel over all nodes. In [30]:

```
from pyspark.mllib.stat import Statistics
dataset= sc.textFile("hdfs://pathtodataset")
dataArray=dataset.map(lambda x:[ array(y) for y in x ])
statistics = Statistics.colStats(dataArray)
print( str(statistics) )
```

Fig. 5. Extracting statistics (Spark/python)

"*Resilient Distributed Datasets (RDDs) are a distributed memory abstraction that lets programmers perform in-memory computations on large clusters in a fault-tolerant manner.* RDDs are typed structures operated in parallel and in-memory. They are immutable (read-only), partitioned into the available nodes (users can specify the number of partitions) and cacheable. RDDs do not mutate, instead operations (transformations) operate on all RDD partitions and generate new RDDs from previous ones. RDDs are stored in-memory as long and as much as possible, and they can be cached if used multiple times. Transformations are evaluated lazily, which means that the data is only transformed when an action, such as showing the results, is submitted. Datasets loaded into an RDD are divided into partitions, and the transformations do associated work on each partition in parallel. The RDD also has a list of parent RDDs (lineage, a graph of transformations where nodes are RDDs). More information on Spark and RDDs is available in [32].

3.1 Shuffling and Other Overheads in Spark

Shuffling is the operation that is executed automatically as part of Spark to re-hash an RDD, re-distributing the data based on those new hash values. More generically, shuffling is the operation of exchanging tuples of a dataset among nodes in any scalability platform. The tuples have to be serialized, sent over the communication channels between nodes and de-serialized again.

Keeping data near the place where it is needed and finding more efficient serialization/de-serialization and compressing approaches can lower the overhead of shuffling. Spark keeps RDDs in-memory as much and as long as possible, but when shuffling is needed or the data no longer fits in-memory serialization and temporary disk storage of parts of datasets may be needed, with disk I/O and garbage collection overheads increasing.

When processing SQL on Spark, joins of large RDDs and other relational operations can generate huge amounts of shuffling. A narrow dependency between successive operations is when the data from a node does not need to be reshuffled into all or most of the other nodes for the next operation, otherwise the dependency between successive operations is called a wide dependency. Wide dependencies generate significant overheads. As in parallel database management systems, these overheads can in principle be reduced or even minimized if data placement and the query optimizer try to maximize equi-partitioning and local processing. Spark also allows users to manually change the partitioning approach and to specify broadcast variables, which are RDDs that are replicated into all nodes. The engine itself, which should analyze the system and decide concerning optimizations, does these kinds of optimizations automatically. For instance, Spark already determines when to broadcast into all nodes during processing.

3.2 Data Frames and Datasets

In spite of offering scalability to data processing, RDD operations, which are specified by developers, cannot take advantage of advanced automated optimization features and efficient execution that is possible with the Data frames and Spark SQL APIs. These allow users to work with structured data, while underneath Spark tries to optimize processing as much as possible using an optimizer - Catalyst – and an execution engine - Tungsten. In the programming API that works directly on RDDs the programmer specifies transformations of RDDs directly, which leaves few optimizations to be done by the query processor. Furthermore, Java serialization/deserialization is used to prepare and retrieve data to be exchanged (shuffled), and memory is allocated from the Java heap. These operations and Java™ garbage collection have associated overheads that can be reduced using more efficient mechanisms. For instance, instead of serializing Java objects, with all the objects overhead, it is possible to create compact binary serializers, and the memory overheads of Java can be avoided using off-the-heap memory. The data frame API statements are submitted to the catalyst query planner optimizer, and execution uses Tungsten, which overrides some of those overheads using off-the-heap shared memory for faster operation and encoders for efficient serialization.

Datasets are a further extension to the data frame API, offering strong typing (type-safe data). Instead of the type row used in data frames, datasets handle data directly in typed structures. This is quite advantageous, since it improves compile time error checking and allows operation directly on user-defined classes. The Datasets API also takes advantage of encoders for converting between JVM objects and compact serialized representations efficiently. The serialized format of datasets is actually a binary tabular data format used by Tungsten, and can be operated on in serialized format efficiently. Encoders for most primitive data types and also for classes are automatically generated and are fast to operate.

The code is quite similar using RDDs or datasets, however datasets have relational operators that ease coding, allow the optimizer to plan for improved execution and use the above described optimizations. The next example confronts word count using RDDs with word count using datasets. The Dataset version uses a count operator that is available in the API (Fig. 6).

The next example shows how to read and process a data file containing Diabetes data into a diabetes dataset defined as diabetes(person:Long, age:Long, sex:String,

On RDDs,

```
val text = sc.textFile("/speech.txt")
val wordCount = text.flatMap(_.split(" ")).filter(_ != "")
  .groupBy(_).map(w => (w._1, w._2.size))
```

On Datasets,

```
val text = sqlContext.read.text("/speech.txt").as[String]
val wordCount = text.flatMap(_.split(" ")).filter(_ != "")
  .groupBy(_).count()
```

Fig. 6. Programming RDDs versus Datasets

region:String, incomeLevel:String, physicalActivity:String, bloodPressure:Double, cardioVascularDesease:String, diabetes:String). After reading, the dataset is queried using a Spark-SQL query (Fig. 7).

```
Val diabetes =
sqlContext.read.text("hdfs:/path/dataset/diabetes.txt").as[String]
val diabetesDS= diabetes.as[(Long, Long, String, String, String,
String, Double, String, String)];
diabetesDS.take(3).foreach(println)
val diabetesGB = sqlContext.sql(
                 "SELECT diabetes, count(*)
                   FROM diabetes
                   GROUP BY diabetes")
```

Fig. 7. Example reading and querying a dataset

Figure 8 shows the phases of query compiling and optimization in Spark-SQL. These are similar to those in traditional relational database management systems, except for the last step that generates RDD processing code corresponding to the operations that are necessary to execute the query. In the figure, the first step constructs an abstract syntax tree (AST) using the tree structure and functionalities of the query compiler, then the AST is analyzed and transformed into a logical query plan (LQP). This LQP goes through logical optimizations to output an "optimized" query plan. The optimized query plan is then translated into multiple possible physical query plans (P1,…,Pn), which represent different possible execution paths to accomplish the same objective.

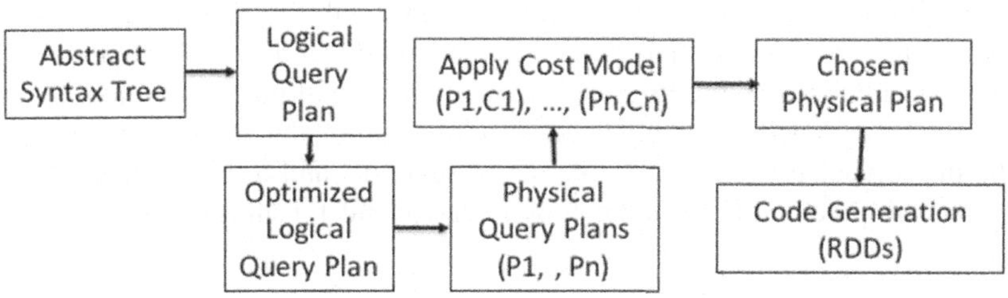

Fig. 8. Catalyst query optimization phases

Given the physical query plans, the next step applies a cost model to each to determine a cost value Ci for each, resulting in pairs (Pi, Ci). The physical query plan with the least cost is then chosen (the one with the smallest cost). The last step of the figure concerns generating the RDD operations that materialize the physical execution plan. The query is then executed by the Spark cluster. In practice, Catalyst will typically not create an "optimal" query plan, since creating "optimal" query plans is too costly and requires too much knowledge of the data set.

Modern approaches to deal with huge amounts of data that were reviewed in this and the previous sections provide flexible and scalable data storage and management for huge amounts of data. Another major concern of big data that needs to be met is the capacity to deal with velocity (realtime). In some contexts, systems need to ingest data at an enormous rate, and/or to process enormous amounts of data while returning answers to queries and to complex analysis quickly. The next section discusses solutions to reach realtime in big data warehousing, and to address realtime scalability limitations of current systems. Other works on Spark and Spark testing while running workloads include [1–3, 21].

4 Realtime Scalable Big Data Analytics

Being able to parallelize and process massive amounts of data does not guarantee realtime, that is, the guarantee that response times are bounded by some value. Yet, many big data applications need or would benefit from realtime capabilities. For instance, recommendation systems could integrate into the recommendation model preferences on new items as soon as people started preferring them, or server intrusion detection systems need to detect intrusions on realtime, and they need to learn from recent accesses.

The major questions that are raised in this context and are the focus of this section are how to design systems that achieve realtime and adjust automatically to guarantee time constraints and scalability regardless of volume, velocity and load. We first introduce the realtime lambda architecture, and then we review some of our recent research work on approaches to provide both realtime and total scalability to data warehousing workloads. The lambda architecture [27] is an abstract architecture designed to contain the components and mechanisms that systems need to deal with big data efficiently, in particular for volume and velocity. That is why it is useful to review this architecture when discussing realtime and total scalability approaches. The approaches that are discuss after reviewing the lambda architecture are a realtime data warehousing architecture based on the addition of a realtime component to the data warehouse, an approach for parallelized de-normalization, an approach using execution tree sharing top scale to massive amounts of sessions, and a mechanism to achieve a single constant predictable response time upper bound for all queries.

4.1 Realtime Concerns

There are two major concerns on realtime analytics: realtime integration of new data and realtime answering of requests (requests include both queries and more complex computations). Traditional approaches for integration of new data dictates that data be integrated at the end of every period, for instance at the end of every day. In contrast, realtime integration demands that data be integrated as soon as it is generated, or shortly after. Similarly, a non-realtime request processing system will receive requests and try to answer them as soon as possible (best effort). Worse still, if many requests are submitted simultaneously, response time will increase significantly. A realtime request processing system ensures that requests are answered as soon as necessary, and regardless of the number of simultaneous queries.

4.2 The Lambda Architecture

The Lambda architecture [27] is an architecture for building realtime big data processing systems. Figure 9 shows the parts involved in the Lambda architecture. Both the batch layer, which stores all data, and the speed layer, which integrates recent data quickly into views, consume new data. The speed layer monitors and integrates data in realtime and reacts immediately on streaming incoming data. It also records and answers low-latency requests, and it refreshes models and views so that they are always up-to-date with the latest data. On the other hand, the batch layer keeps all the data in immutable format. There are no deletes, updates or changes to the data in that layer. This way the data in the batch layer is always complete and can be used to reconstruct any model or view. The size of the batch layer can be huge, therefore pre-processed views are used to avoid re-running all queries.

The service layer contains code and functionality to answer client requests, using the data and models from the batch and speed layers.

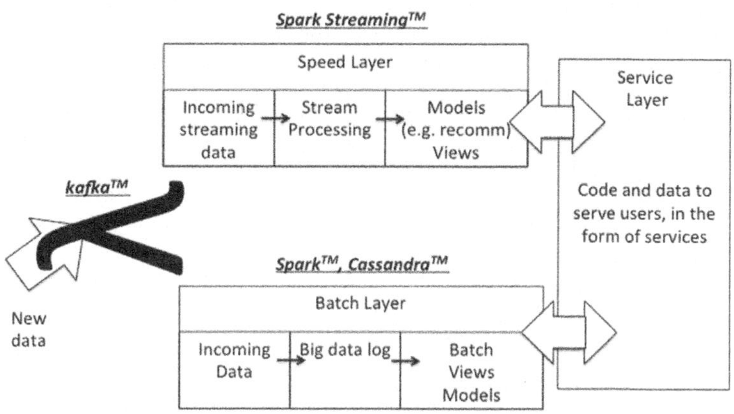

Fig. 9. Overview of lambda architecture

Figure 9 also shows a possible assignment of Spark and Hadoop-related technologies to handle parts of the Lambda architecture. A scalable publish-subscribe messaging software such as Kafka™ [36] can implement a producer/consumer pattern for incoming data. The batch layer can use Spark™ on Hadoop™ or other scalable platform, and some NoSQL data store such as Cassandra™, while the speed layer can be implemented using Spark Streaming™ [35] programmed and integrated with Spark itself. Spark Streaming is an extension to Spark that applies the same Spark API to stream processing, for creating streaming applications using the same environment and functions you would use for batch jobs in Spark™. It integrates completely in Spark and is actually part of Spark eco-system.

4.3 Realtime Data Integration

In [13] we describe the details of an approach to add realtime data integration and querying to a data warehouse architecture. The approach adds a separate realtime component to the data warehouse. When compared with the lambda architecture, the realtime component is a possible realization of the speed layer, and it can be added to a traditional data warehouse architecture to handle velocity.

The main parts of the system are shown in Fig. 10 The realtime component is assumed to hold the last period of data, e.g. the current day. The data warehouse is assumed to be made of a set of star schemas, materialized views and auxiliary structures (indexes and so on). The realtime component has the same schema as the data warehouse, which allows the transformation and loading to work as before. The main difference is that now new data can go through the whole transformation and loading steps as soon as it appears (row-by-row) or in small batches. Transformation and loading can be tuned for row-by-row or small batch processing, to take advantage of the realtime component. The realtime component holds the most recent data that is not yet integrated into the main data warehouse. It should be in a separate machine or node from that of the main data warehouse. This way loading and querying the realtime component will not affect the speed of query processing over the main data warehouse. The realtime component should work solely in-memory if possible, and it can be

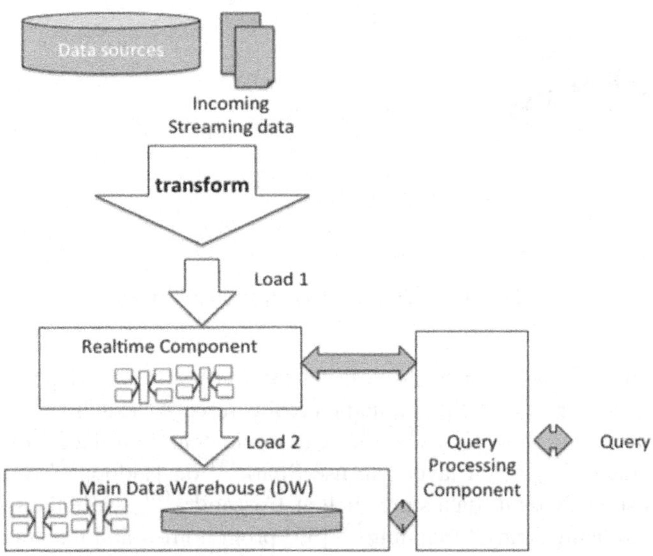

Fig. 10. Realtime data warehouse

implemented with an in-memory database. It can include the same materialized views and auxiliary structures as the main data warehouse, however that may be unnecessary, at least partly, since it has to deal with much smaller datasets.

On top of Fig. 10 we can see the data coming into the realtime component. The load defined as load 2 is a periodical loading of data from the realtime component to the main data warehouse. As soon as all data is loaded into the main data warehouse, the realtime component is emptied again and starts working with new data only. This architecture still requires periodical loading of the main data warehouse, but it ensures realtime integration of new data, and eliminates transformation during the loading of the main data warehouse, since transformation is done while integrating the new data into the realtime component.

Figure 10 also shows a query processing component. This component needs to distinguish between queries to be executed on the data warehouse, those to be executed on the realtime component, and those to be ran against both. The user may specify which case he wants using a simple extension to the SQL query syntax, or the query processing component can analyse automatically the time-date intervals of the submitted query to determine the target. If the target is both components, the query processing component re-writes the query to generate a query for each component and a merge query that merges the results from both components. This process uses the same algorithms as the parallel query processing algorithms in [17, 19] and all the operations are implemented transparently to the user.

To show the improvement resulting from the realtime data warehouse architecture, we created a prototype and ran the setup of [13] using a modified Star Schema Benchmark (SSB) (schema and data) [28] as the data warehouse, and TPC-H update sets as the new data to be loaded. The SSB star schema benchmark was modified to make it capable of evaluating realtime capabilities, calling the resulting setup SSB-RT. The data model for this setup is made of two stars: (A) the sales star (Orders), and dimension tables (Time_dim, Date_dim e Customer); (B) a star holding lineitem, with five dimension tables (Time_dim, Date_dim, Customer, Suppllier and Part). There are also 12 materialized views measuring sales per supplier, part, customer against granularities (hour, day, month and year). There are two auxiliary tables for computation of refresh sets for the materialized views, one for each fact. In order to load these tables initially, the DBGEN utility of TPC-H was used, creating the text files with information from the data sources. Those files were transformed into the star schema model, simulating the data extraction and transformation process of a data warehouse. After the transformation, the data was loaded into the data warehouse. The initial database size was 10 GB.

The realtime component had the same schema as the main data warehouse. For refreshing the data warehouse with new data (Extract, Transform, Load and Refresh, we assumed that the information to be loaded was extracted from the data sources and stored in log files. The SSB-RT reads the logs and runs the transformation operations (T) before the loading (L) of the stars into the data warehouse. When using the realtime component, the load was done to an in-memory database instead of the main data warehouse.

The experiments used three servers: one for the staging area, doing transformations, one for the realtime component and one for the data warehouse itself. The staging area server was an Intel(R) Core(TM) i7 1,80 GHz with 8 GB memory. The realtime component was ran on an Intel(R) Core(TM) 2 Quad 2,5 GHz and 4 GB of RAM, while the data warehouse used an Intel(R) Core(TM) i5 3,40 GHz with 16 GB of RAM. In order to get statistically significant data, we ran each case 15 times, then computed the average response times and discarded the 5 most distant results from the average, then recalculating the average.

The prototype uses an industry-strength database management system to store and process the data warehouse, plus the realtime component built on an in-memory DBMS and holding up to one-day's worth of data. It implements query re-writing and merging to allow queries to be executed against both components (realtime and main data warehouse). The conclusion from the experiment was that the realtime setup was able to do realtime loading while querying without any significant delays in either loading or query answering, as shown in Fig. 11. The figure shows query performance for SSB queries when ran against a traditionally organized data warehouse, versus when executed on the improved architecture. The queries are SSB queries (q1 = Q1.1, q2 = Q1.2, q3 = Q1.3, q4 = Q2.1, q5 = Q2.2, q6 = Q2.3, q7 = Q3.1, q8 = Q3.2, q9 = Q3.3, q10 = Q3.4, q11 = Q4.1, q12 = Q4.2, q13 = Q4.3). The first column of results for each query shows the response time when running queries, one at a time. The second column shows the response time when we add simultaneous loading and refreshing of data warehouse structures in a traditional architecture. The third column shows the results obtained when we added the realtime component with data loading and refreshing simultaneous to query processing. While the performance degraded significantly in the traditional data warehouse, when loading and querying occurred simultaneously, response times while loading are almost the same with the realtime component as they were in the traditional data warehouse architecture without loading any data.

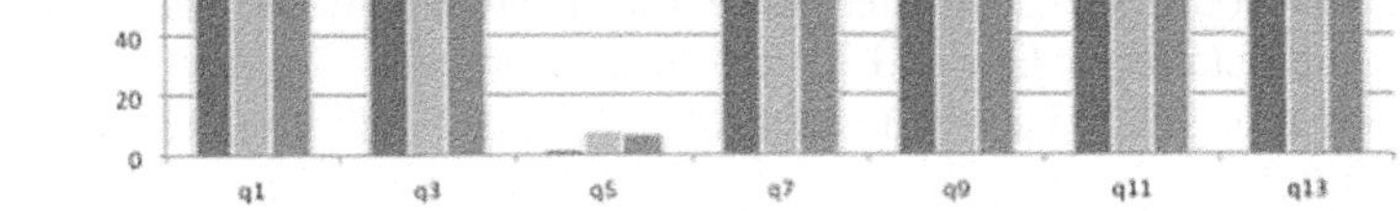

Fig. 11. Realtime data warehouse architecture

In [13] we tested many other details to conclude that the realtime component enables realtime integration into the data warehouse, without affecting query processing significantly. More details on the realtime data warehousing approach and other alternatives are available in [12, 13, 29]. An approach for totally scalable and realtime data warehouse that adds automated scalability to this setup is discussed in detail in references [22–26].

4.4 De-normalization and Predictable Response Time

It is not possible to predict tight execution time bounds for queries in Relational DBMSs, due to the complexity of query processing. Relational DBMSs rely on powerful query optimizers that plan complex operation trees to execute queries as efficiently as possible. The query optimizer must build a logical query plan, improve that plan, generate multiple possible physical query plans, evaluate the costs of each physical plan and choose the one with the least cost. During the planning, there are alternative orders for scanning and joining tables or intermediate datasets. Some operations such as sorts or hashing need to be applied over huge datasets. Large intermediate result sets are frequently generated, and those may have to be paged when memory is not enough. There are alternative access methods, structures such as indexes (b-tree, bitmap, others) and materialized views, multiple possible algorithms for some operations (e.g. a join may be done using hash-join, sort-merge-join or nested loops, with or without indexes in some steps), and there are also sometimes more sophisticated algorithms and on-the-fly structures such as bitmaps to speedup processing.

It is much easier to predict an execution time bound if we have a single large table or relation and queries that access only that relation, since operation is straightforward. Consider a data warehouse star schema, made of a central fact table and several dimension tables that are referenced by the fact table. In [4, 5, 11] we have explored an approach, and built a middleware that, at a physical level, de-normalizes the schema into a single relation that contains facts and all the associated dimensions information. The approach is illustrated in Fig. 12.

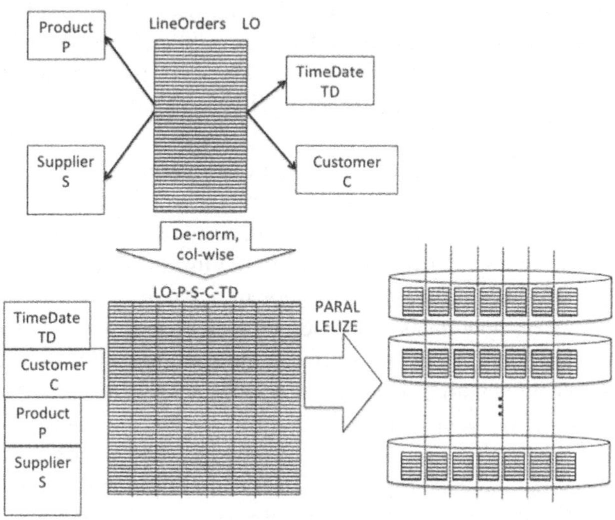

Fig. 12. Transformations: Star -> De-normalized -> Parallel

The de-normalized schema contains massive redundancy of dimension data and can be stored as compressed column-wise data for faster access to individual columns.

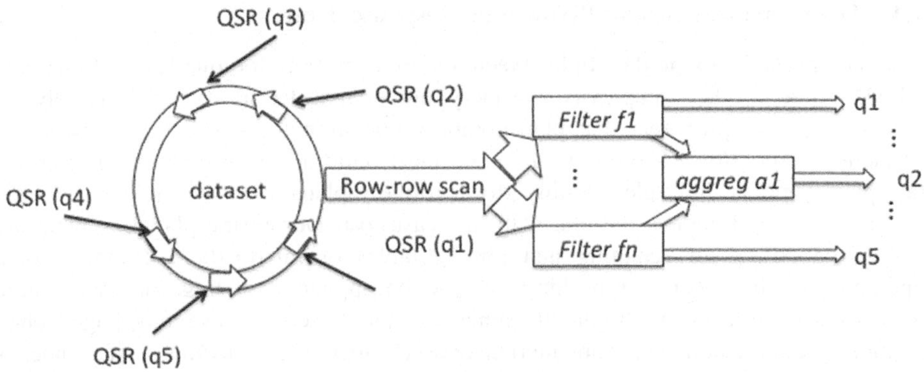

Fig. 13. SPIN execution sharing

The first step of Fig. 12 shows the de-normalization and storage as column-wise. Although this approach results in a big relation (in experiments with TPC-H [38] the data size increased around 3.5 to 4 times), there are a set of important advantages. The first is that it becomes trivial to parallelize into multiple nodes (by dividing rows into the number of available nodes) and each node will be able to process the data mostly independently from other nodes. The second step of Fig. 13 is parallelization. Query processing becomes trivial over the de-normalized schema, and it also avoids most shuffling in the parallel de-normalized version. There are no complex joins (all the data is already pre-joined into the single relation), and all query processing is based on a single scan of the whole dataset, or parallel scans in the parallel version. Most relational operations are straightforward to implement, in particular filters are applied as rows are being scanned from the single dataset. Aggregations and group by are implemented by hash lookups during the scan. For more details on query processing under these organizations, please refer to [4, 5, 9–11].

The de-normalized data is always appended in read-only manner (immutable). In this aspect, it is similar to the Batch Layer of Lambda architectures, and it is a log of all data. Dimension data in a row of the de-normalized relation represents the view of the corresponding dimension values as they were in the instant when the fact was captured. The current version of the dimensions is kept in separate dimension relations, to be used on integration of new facts into the de-normalized schema and for querying current values (e.g. a query accessing only a dimension does not need to be processed over the de-normalized schema, instead it scans the dimension relation).

In spite of its advantages, this approach requires rethinking query optimization and processing, which are out of the scope of this work. The approach is motivated by simpler cost estimation and easier join execution, and in particular the capacity to scale easily to any number of nodes. However, we already identified the problem of increase in data size as one drawback. The approach also requires re-thinking how data is operated upon. For instance, consider a change in a customer address. Since customer is a dimension, and the value of that row in the dimension changes, how do we update the de-normalized dataset to reflect the new address? First of all, we update the address information in the dimension table. But the customer information also appears in every row of the de-normalized table that refers to that customer (e.g. every sale to that customer). Does the query processing engine need to change every occurrence of the address for that customer, which would be quite heavy? The answer is no, the address that is in every row of the dataset referencing that customer is correct, since it represents his address at the date and time corresponding to that fact, and facts added to the de-normalized table after the change in address will have the new address. Another case is self-joins. The question is, how to implement a self-join efficiently in the de-normalized dataset? Like in traditional relational database management systems, the answer to this question has to do with auxiliary structures (indexes) and with the query optimizer, which has to be designed for the modified model. Please, refer to [4, 5, 9–11] for more details on the approach and related mechanisms.

4.5 Session Scalability

In the context of Relational DBMSs, the term "one-query-at-a-time model" refers to the fact that, when a query is submitted, the query processor creates an execution plan for that query that is independent from those of other queries that might be running, and all the processing done by the execution plan is independent of the processing for other queries. If 50 queries are running simultaneously, each will have to process everything on its own. Each query needs to scan each row of tables and indexes it uses, each query has to join and aggregate on its own and to produce intermediate and final results. Each session needs memory for sometimes-huge intermediate results and for paging into intermediate files when memory is not enough. This, together with the complex processing algorithms and structures, results in a system exhibiting significant slowdown as soon as more sessions are executing queries simultaneously.

In [5–8] Joao Costa et. al. propose a different model for use with the de-normalized data organization discussed in the previous section. The SPIN model uses execution tree sharing between queries that execute on the data organized as a single de-normalized table. Recall that query processing over the de-normalized data is much simpler than in traditional relational organization, because there are no joins and all processing is based on scanning, filtering and aggregating over the single existing table. In the new model, the root of the execution tree is the scan of the de-normalized table. All queries share the same scanning of the table, which is scanned completely, one row at a time, the rows being fed to operators in the execution tree (if the table is organized column-wise, then each column that is needed is scanned by a separate scanner thread). The approach is called SPIN because the scanning is repeated cyclically (like a disk spinning) while there are still queries being processed.

When a query is submitted, the execution plan is created and "plugged" into the shared query-processing tree. The query starts consuming rows in the point of the scanning that corresponds to the row being scanned currently (query starting row QSR), and then it will only stop consuming rows when the scanning completed a whole cycle of scanning rows, from the QSR to the last row of the table and back from the first row of the table to the QSR. Figure 13 illustrates the processing. In the Figure, queries q1 to q5 start consuming rows in different points of the scanning of the table (dataset), as shown by the QSR arrows in the diagram. Each of those queries will only complete its processing after the cyclic scanning does a whole cycle and gets back to the QSR point where the query started. Each row being scanned is fed to the pipeline in the right of the figure. Filters and other operators, such as aggregation, are placed in the shared query execution tree and, when possible, shared between queries. Then each query completes its more specific processing within its own query processor thread. In the query execution tree shown in the right of the figure, blocks to the right are consumers of data produced by the blocks immediately to the left of them, and all operations are shared as much as possible. For detailed information on how operations such as aggregations, sub-queries and other operation blocks are executed, please refer to [5–8].

It is also easy to understand how the parallel version of SPIN works. Each node will have the execution tree shown in Fig. 13, but scanning only the *1/n* of the data that resides in the node. In the parallel case, this execution tree computes per-node partial results, which need to be merged in some node(s) in a global execution tree that includes these partial execution trees for each node. An on-demand execution approach can also be added to this architecture, by dividing the dataset into chunks and load-balancing execution.

4.6 Scaling for Approximate Execution Time Bound Guarantees

As we already described in a previous section, execution of a typical OLAP query in the de-normalized architecture involves a single scan of the whole table and simple in-memory operations to filter, aggregate and so on. We observed that, while the execution times in a traditional Relational DBMSs are heterogeneous (e.g. different TPC-H queries take very distinct times to complete), increase dramatically when multiple sessions submit queries simultaneously, and depend a lot on the resources available (e.g. paging of intermediate results will slow down processing significantly), the same queries take an almost constant time to complete in the de-normalized organization and query processing mechanisms (note that subqueries count as additional queries in this thinking, and queries accessing only dimensions are processed quickly, against the dimension tables). For this reason, it becomes feasible in parallel SPIN to predict an approximate value for the execution time of queries, which depends mostly on IO for scanning the table. For a read rate of *r* MB/sec, a dataset of size S is scanned in approximately *S/r* seconds. If we assume that the remaining in-memory operations necessary to complete processing the queries adds an overhead of *o%*, the approximate estimation of the query runtime is given by Eq. (1),

$$t = S/r * (1 + o)\, seconds \tag{1}$$

In the case of parallel placement and operation of the de-normalized schema on *n* nodes, the independent query processing of dataset parts by nodes is estimated to take the time shown in Eq. (2)

$$t = S/r * (1 + o)/n\, seconds \tag{2}$$

To this processing time, it is necessary to add the time taken to do data exchange and to apply a merge query. Based on these predictions, it is possible to create a scalable parallel data management system that scales to meet maximum query execution time bounds. In [9, 10] Joao Costa et al. propose a middleware (Teepa) that implements this capability on the parallel, de-normalized schema. In Teepa, administrators can specify desired maximum execution time bounds for queries. Given Eq. (2), if we setup n nodes as determined by Eq. (3), we will be able to approximately meet the maximum execution time constraints T when answering queries.

$$n = S/r * (1 + o)/T \, nodes \tag{3}$$

For more information on this kind of approach, please refer to the works on TEEPA [9, 10] and the Phd thesis in [5].

5 Conclusions

In this work we discussed models, architectures, concepts and technologies related to scalability and realtime in big data analytics using modern big data platforms. We also discussed query processing performance issues, and discussed our own recent research efforts to create improved solutions. Our presentation included the major scalability architectures, from parallel database systems to NoSQL, MapReduce and the lambda architecture, some technologies (Spark, Spark-SQL and Spark echo-system technologies). The research proposals discussed were approaches to overcome current limitations in terms of realtime and automated scalability of the entire big data warehousing pipeline. Overall, we focused more on concepts than specific technologies, since concepts remain relevant for a significant time frame, while new and improved technologies will surely quickly replace current technologies.

References

1. Armbrust, M., Xin, R.S., Lian, C., Huai, Y., Liu, D., Bradley, J.K., Meng, X., Kaftan, T., Franklin, M.J., Ghodsi, A. and Zaharia, M.: Spark SQL: relational data processing in spark. In: Proceedings of the 2015 ACM SIGMOD International Conference on Management of Data, SIGMOD 2015 (2015)
2. Chiba, T., Onodera, T.: Workload characterization and optimization of TPC-H queries on Apache Spark, Research Report RT0968, IBM Research – Tokyo, 16 October 2015
3. Chiba, T., Onodera, T.: Workload characterization and optimization of TPC-H queries on Apache Spark. In: 2016 IEEE International Symposium on Performance Analysis of Systems and Software (ISPASS) (2016)
4. Costa, J., Cecílio, J., Martins, P., Furtado, P.: ONE: predictable and scalable DW model. In: International Conference on Big Data Analytics and Knowledge Discovery (2011)
5. Costa, J.P.: Massively scalable data warehouses with performance predictability, PhD thesis, University of Coimbra, July 2015

6. Costa, J.P., Furtado, P.: Data warehouse processing scale-up for massive concurrent queries with SPIN. In: Hameurlain, A., Küng, J., Wagner, R., Bellatreche, L., Mohania, M. (eds.) Transactions on Large-Scale Data- and Knowledge-Centered Systems XVII. LNCS, vol. 8970, pp. 1–23. Springer, Heidelberg (2015). doi:10.1007/978-3-662-46335-2_1
7. Costa, J.P., Furtado, P.: Improving the processing of DW star-queries under concurrent query workloads. In: Bellatreche, L., Mohania, Mukesh K. (eds.) DaWaK 2014. LNCS, vol. 8646, pp. 245–253. Springer, Cham (2014). doi:10.1007/978-3-319-10160-6_22
8. Costa, J.P., Furtado, P.: SPIN: concurrent workload scaling over data warehouses. In: Bellatreche, L., Mohania, Mukesh K. (eds.) DaWaK 2013. LNCS, vol. 8057, pp. 60–71. Springer, Heidelberg (2013). doi:10.1007/978-3-642-40131-2_6
9. Costa, J.P., Martins, P., Cecilio, J., Furtado, P.: Providing timely results with an elastic parallel DW. In: Chen, L., Felfernig, A., Liu, J., Raś, Z.W. (eds.) ISMIS 2012. LNCS, vol. 7661, pp. 415–424. Springer, Heidelberg (2012)
10. Costa, J.P., Martins, P., Cecilio, J., Furtado, P.: TEEPA: a timely-aware elastic parallel architecture. In: Proceedings of the 16th International Database Engineering & Applications Symposium, IDEAS 2012, Prague, Czech Republic (2012)
11. Costa, J.P., Cecílio, J., Martins, P., Furtado, P.: Overcoming the scalability limitations of parallel star schema data warehouses. In: Xiang, Y., Stojmenovic, I., Apduhan, B.O., Wang, G., Nakano, K., Zomaya, A. (eds.) ICA3PP 2012. LNCS, vol. 7439, pp. 473–486. Springer, Heidelberg (2012). doi:10.1007/978-3-642-33078-0_34
12. Ferreira, N., Furtado, P.: Near real-time with traditional data warehouse architectures: factors and how-to. In: 17th International Database Engineering and Applications Symposium (2013)
13. Ferreira, N., Furtado, P.: Real-time data warehouse: a solution and evaluation. Int. J. Bus. Intell. Data Min. **8**(3), 244–263 (2014)
14. Furtado, P.: Experimental evidence on partitioning in parallel data warehouses. In: Proceedings of the ACM DOLAP 2004 - Workshop of the International Conference on Information and Knowledge Management, Washington USA, November 2004
15. Furtado, P.: Workload-based placement and join processing in node-partitioned data warehouses. In: Proceedings of the International Conference on Data Warehousing and Knowledge Discovery, Zaragoza, Spain, pp. 38–47, September 2004
16. Furtado, P.: Efficient and robust node-partitioned data warehouses. In: Wrembel, R., Koncilia, C. (eds.) Data Warehouses and OLAP: Concepts, Architectures and Solutions, Chap. IX, pp. 203–229. Ideas Group, Inc. ISBN 1-59904365-3
17. Furtado, P.: A survey of parallel and distributed data warehouses. Int. J. Data Warehous. Min. **5**(2), 57 (2009)
18. Furtado, P.: Replication in node-partitioned data warehouses. In: DDIDR2005 Workshop of International Conference on Very Large Databases, VLDB (2005)
19. Furtado, P.: Efficiently processing query-intensive databases over a non-dedicated local network. In: Proceedings of the 19th International Parallel and Distributed Processing Symposium, Denver, Colorado, USA, May 2005
20. Furtado, P.: Model and procedure for performance and availability-wise parallel warehouses. Distrib. Parallel Databases **25**(1), 71 (2009)
21. Furtado, P.: Scalability and Realtime for Data Warehouses and Big data, Paperback, 11 September 2015
22. Martins, P.: Elastic ETL+Q for any data-warehouse using time bounds. PhD thesis, University of Coimbra, February 2016
23. Martins, P., Abbasi, M., Furtado, P.: Data-warehouse ETL+Q auto-scale framework. Int. J. Bus. Intell. Syst. Eng. **1**(1), 49–76 (2015)

24. Martins, P., Abbasi, M., Furtado, P.: AutoScale: automatic ETL scale process. In: 19th East European Conference on Advances in Databases and Information Systems (2015)
25. Martins, P., Abbasi, M., Furtado, P.: Preparing a full auto- scale framework for data-warehouse ETL+Q. In: IEEE Big data Congress 2015, New York, USA (2015)
26. Martins, P., Abbasi, M., Furtado, P.: AScale: automatically scaling the ETL+Q process for performance. Int. J. Bus. Process Integr. Manage. **7**(4), 300–313 (2015)
27. Marz, N., Warren, J.: Big Data: principles and best practices of scalable realtime data systems, 1st Manning Publications Co. Greenwich, CT, USA ©2015 (2015), ISBN:1617290343 9781617290343
28. O'Neil, P., O'Neil, E., Chen, X.: Star schema benchmark - revision 3. Technical report, UMass/Boston (2009)
29. Waas, F., Wrembel, R., Freudenreich, T., Thiele, M., Koncilia, C., Furtado, P.: On-demand ELT architecture for right-time BI: extending the vision. Int. J. Data Warehous. Mining **9**(2), 21–38 (2013)
30. Zaharia, M., Chowdhury, M., Das, T., Dave, A., Ma, J., McCauley, M., Franklin, M.J., Shenker, S., Stoica, I.: Resilient distributed datasets: a fault-tolerant abstraction for in-memory cluster computing. In: Proceedings of the 9th USENIX Conference on Networked Systems Design and Implementation (NSDI 2012). USENIX Association, Berkeley, CA, USA, p. 2 (2012)
31. Zhao, J., Pjesivac-Grbovic, J.: MapReduce, the programming model and practice. In: Sigmetrics/Performance 2009, Tutorials, 19 June 2009
32. Spark Homepage. http://spark.apache.org/. Accessed Jul 2016
33. Spark SQL homepage. http://spark.apache.org/sql/. Accessed Jul 2016
34. Parquet File Format. https://parquet.apache.org/. Accessed Jul 2016
35. Spark Streaming. http://spark.apache.org/streaming/. Accessed Jul 2016
36. Kafka homepage http://kafka.apache.org/. Accessed Jul 2016
37. CassandraTM Homepage. http://cassandra.apache.org/. Accessed Jul 2016
38. TCP Council homepage. www.tpc.org. Accessed Jul 2016
39. Snijders, C., Matzat, U., Reips, U.-D.: 'Big Data': Big gaps of knowledge in the field of Internet. Int. J. Internet Sci. **7**, 1–5 (2012)
40. Hashem, I.A.T., Yaqoob, I., Anuar, N.B., Mokhtar, S., Gani, A., Khan, S.U.: The rise big data on cloud computing: review and open research issues. Information Systems **47**, 98–115 (2015)
41. "Data, data everywhere". The Economist. 25 February 2010. Retrieved 2 December 2016
42. "Supercomputing the Climate: NASA's Big Data Mission". CSC World. Computer Sciences Corporation. Retrieved 2 December 2016
43. "DNAstack tackles massive, complex DNA datasets with Google Genomics". Google Cloud Platform. Retrieved 20 December 2016
44. Mirkes, E.M., Coats, T.J., Levesley, J., Gorban, A.N.: Handling missing data in large healthcare dataset: a case study of unknown trauma outcomes. Comput. Biol. Med. **75**, 203–216 (2016)
45. Brewer, E.: CAP twelve years later: how the "rules" have changed. Computer **45**(2), 23–29 (2012)
46. Chang, F., Dean, J., Ghemawat, S., Hsieh, W.C., Wallach, D.A., Burrows, M., Chandra, T., Fikes, A., Gruber, R.E.: Bigtable: a distributed storage system for structured data, OSDI 2006: Seventh Symposium on Operating System Design and Implementation, Seattle, WA, November 2006
47. Google File System and BigTable, Radar (World Wide Web log), Database War Stories (7). O'Reilly, May 2006

48. Angles, R., Gutierrez, C.: Survey of graph database models. ACM Comput. Surv. **40**(1), 1 (2008)
49. Robinson, I., Webber, J., Eifrem, E.: Graph Databases. 2nd edn. O'Reilly Media (2015)
50. Cattell, R.: Scalable SQL and NoSQL data stores. SIGMOD Rec. **39**(4), 12–27 (2010)
51. Han, J., et al.: Survey on NoSQL database. In: 2011 6th International Conference on Pervasive Computing and Applications (ICPCA). IEEE (2011)
52. Grolinger, K., et al.: Data management in cloud environments: NoSQL and NewSQL data stores. J. Cloud Comput. Adv. Syst. Appl. **2**(1), 1 (2013)
53. Graefe, G.: Volcano - an extensible and parallel query evaluation system. IEEE Trans. Knowl. Data Eng. **6**(1), 120–135 (1994)
54. Kossmann, D.: Distributed query processing approaches. In: ACM Computing Surveys (CSUR) (2000)
55. Deshpande, A., Ives, Z., Raman, V.: Adaptive query processing. Found. Trends Databases **1**(1), 1–140 (2007)
56. Dean, J., Ghemawat, S.: MapReduce: simplified data processing on large clusters. Commun. ACM **51**(1), 107–113 (2008)
57. Lakshman, A., Malik, P.: Cassandra: a decentralized structured storage system. ACM SIGOPS Oper. Syst. Rev. **44**(2), 35–40 (2010)

Step by Step Towards Energy-Aware Data Warehouse Design

Ladjel Bellatreche[1(✉)], Amine Roukh[2], and Selma Bouarar[1]

[1] LIAS/ISAE-ENSMA - Poitiers University, Poitiers, France
{bellatreche,bouarars}@ensma.fr
[2] University of Mostaganem, Mostaganem, Algeria
roukh.amine@univ-mosta.dz

Abstract. Nowadays, the electrical energy efficiency is one of the most challenging issues in the area of ITs. DBMS have been pointed out as one of the major energy consumers. The reduction of their energy consumption becomes an urgent priority. Two aspects have to be considered in order to reduce this consumption: **(i)** the DBMS hosting the database applications and **(ii)** the Eco-design of these applications. Note that the first aspect got more attention than the second one. In this paper, we attempt to consider both aspects in the context of data warehouses ($\mathcal{DW}$). Firstly, we propose a generic framework integrating the energy in query optimizers of DBMS hosting already designed $\mathcal{DW}$. An instantiation of this framework has been done on PostgreSQL DBMS. Secondly, and thanks to the variability that has been widely studied by the community of software, we propose to go back to the logical phase of the $\mathcal{DW}$ life cycle and see how the energy may be integrated and then evaluate its impact on the physical phase. This variation is possible due to the *relationships* (*hierarchies*) that may exist among the properties of that $\mathcal{DW}$. Finally, intensive experiments are conducted to evaluate the effectiveness and efficiency of our findings on PostgreSQL and Oracle DBMS.

1 Introduction

The COP21 (United nations conference on climate change)[1] event has shown the willingness of countries (over 145 foreign Heads of State and Government attended the conference at Le Bourget, Paris, France), companies[2], individuals, government and non-government associations[3], etc. to save the planet. In high developed countries, the buildings consume about 40% of the total national energy use. This contributes in increasing greenhouse gas emissions. Computers and other electronic devices account for about 6% of all building energy use.

[1] http://www.gouvernement.fr/en/cop21.
[2] http://climateaction.unfccc.int/assets/downloads/LPAA_-_Private_sector_engagement.pdf.
[3] http://www.energyforall.info/.

P. Marcel and E. Zimányi (Eds.): eBISS 2016, LNBIP 280, pp. 105–138, 2017.
DOI: 10.1007/978-3-319-61164-8_5

The U.S. Energy Information Administration forecasts that energy use in data center servers will increase five-fold by 2040[4].

Several efforts and initiatives have been launched to reduce the electrical energy consumption of private and public buildings. For instance, the *Australian government* has established an energy efficiency strategy since 2002[5]. This strategy describes the measures that have been taken in terms of energy management within government buildings. These measures intervene at *each stage of a building's life*: **(a)** design and construction of new government buildings; **(b)** refurbishment of existing government buildings; **(c)** negotiation of new and renewed leases; **(d)** ongoing building management and maintenance; and **(e)** equipment procurement.

The community of building is already ahead of the curve in energy saving unlike the $\mathcal{DB}$ community that has been pointed out as one of the major energy consumers. In typical data centers, *DBMS* is *one of the most important consumers of computational resources among other software deployed*, which turns DBMS to be a considerable energy consumer [33]. Reducing the electrical energy consumption in the context of databases ($\mathcal{DB}$) represents a crucial issue. The *Beckman report* on databases published in February 2016 has already highlighted this issue [1]. Face to this worrying situation, the $\mathcal{DB}$ community did not stand idly, but from the last decade, it continuously proposes initiatives centred on four main actions (that we call **OBRE**: **O**ffer, **B**orrow, **R**eform, **E**valuate) to deal with energy. The particularity of these actions is that they consider DBMS hosting already designed databases.

- **Offer:** The database technology was made available for energy professionals to the analysis purpose enabling smarter scheduling of energy consumption of entities such as smart cities (in MIRABEL project [42]) and electrical vehicles in EDF (*Electricité De France*) project [40].
- **Borrow:** The $\mathcal{DB}$ technology employs green hardware, platforms and techniques to deploy the target database applications [20]. Several collaborations between database editors and hardware companies have been developed. We can cite the collaboration between *Oracle Exadata Database Machine* [20] and *Intel Corporation* to reduce the energy costs. Other techniques such as the deactivation of electronic component during periods of CPU inactivity and the usage of *Dynamic Voltage and Frequency Scaling* technique aiming at adjusting on-the-fly the voltages and frequencies of CPU [23].
- **Reform:** A couple of efforts have revisited the $\mathcal{DB}$ software in order to make them more energy sensitive. They mainly concern the query optimizes, where mathematical cost models, estimating the energy consumption when executing queries, have been developed and used to: (i) generate query plans [24,25,48] and (ii) select optimization structures such as materialized

[4] https://energy.gov/sites/prod/files/2015/09/f26/Quadrennial-Technology-Review-2015_0.pdf.

[5] https://www.sa.gov.au/topics/property-and-land/land-and-property-development/building-rules-regulations-and-information/sustainability-and-efficiency-regulations/government-energy-efficiency-initiatives.

views [38]. It should be noticed that these efforts focus on DBMS hosting the target $\mathcal{DB}/\mathcal{DW}$ and do not focus on integrating energy during its design process.
- **Evaluate:** to test their efficiency and effectiveness, energy initiatives have to be evaluated using either real or synthetic datasets [38].

In this paper, we only focus on electrical energy savings in databases from software perspective. Our solutions are split into two directions:

(i) The consolidation of the initial efforts on making query optimizers more energy sensitive [24,25,38,39,48]. There have been extensive studies on the query optimization, since the early 70s. Note that advanced query optimizers perform two main tasks: **(1)** enumeration of execution plans of a given query and **(2)** the selection of the best plan. This selection uses cost-based optimization (CBO). The CBO is a mathematical processor, where it uses formulas to evaluate the cost of a plan. This cost is dedicated to estimate several non-functional requirements (NFR) such as the number of inputs-outputs, CPU, communication, etc. A CBO driven-approach is suitable when statistics on tables, indexes, selectivity factors of join and selection predicates, etc. are available. A straightforward way to integrate the electrical energy in a query optimizer is to develop new mathematical cost models to estimate it during queries execution. In this paper, we will present a comprehensive framework on how to build Eco-query optimizers and we instantiate it in the context of the *PostgreSQL* DBMS, considered as the world's *most advanced open source database.*

(ii) The integration of the energy dimension in the most important phases of $\mathcal{DW}$ life cycle composed of the following phases: **(a)** data source analysis, **(b)** elicitation of requirements, **(c)** conceptual, **(d)** logical, **(e)** deployment, and **(f)** physical design. ETL is appended as a design phase responsible for analytical processing [21] (Fig. 1).

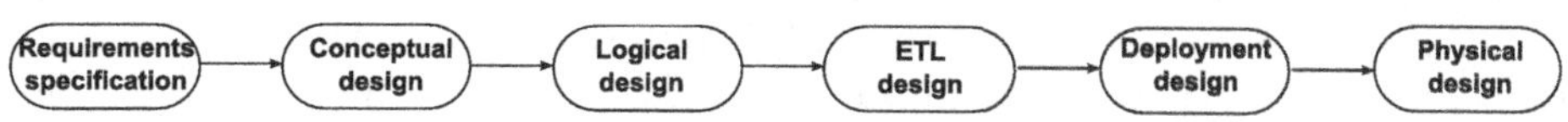

Fig. 1. Phases of $\mathcal{DW}$ design

To the best of our knowledge, this is the first work that shows how the electrical energy could be integrated along the life cycle of the $\mathcal{DB}/\mathcal{DW}$. It should be noticed that existing studies dealing with Eco-query optimizers consider a $\mathcal{DB}/\mathcal{DW}$ with a frozen logical model. This assumption surely becomes questionable when integrating energy in the design, since it ignores the chained aspect of the life cycle. Knowing that many variants of a logical schema may exist. This is due to the presence of dependencies and hierarchies among attributes of a given warehouse application. As a consequence, it is worth studying the impact of this variation on the energy consumption.

The paper is organized as follows: Sect. 2 presents a discussion about the existing works on energy based on the four actions that we have mentioned in the Introduction Section which are **O**ffer, **B**orrow, **R**eform and **E**valuate. Section 3 gives some definitions about energy. Section 4 presents our first contribution in integrating the energy in query optimizers. Section 5 discusses the integration of the energy in two phases of the life cycle of $\mathcal{DB}/\mathcal{DW}$ design representing the logical and the physical steps. Finally, Sect. 6 concludes the paper by summarizing the main results and suggesting future work.

2 Related Work

In this section, we discuss the different initiatives and efforts made by the database community according to *OBRE* actions.

Offer. As we said in the Introduction, the database technology is used in the past and now to store energetic data from vehicles, smart cities, etc. The *MIRABEL Project* [42] is an example of this direction. It consists in developing an approach on a conceptual and an infrastructural level that allows energy distribution companies balancing the available supply of renewable energy sources and the current demand in ad-hoc fashion. It uses a DBMS to store forecasting in order to answer time series queries, which is an important functionality in energy data management [15,30].

Borrow. As other technology, databases never stop borrowing green hardware and platforms for their applications and DBMS. These works span from processing devices such as CPU [7,20,26,49], FPGA [46], GPU [36], CPU-GPUs co-processing [19] to storage management such as repartitioning the database [17], data migration and disk state adjustment [6,44], using SSDs as a storage device instead of HDD [14,41]. Moreover, recent studies such as [5] highlights that the growing use of main memory databases will soon emerge memory as the dominant power consumer instead of CPU. Based on that, the authors used DRAM frequency scaling and power-down modes to improve power consumption. A similar approach has be used in [23]. The work of [18] proposed a hybrid memory architecture consisting of both DRAM and Non-Volatile Memory (NVM) to reduce database energy consumption. However, in this study, we argue that hardware-only approaches are only part of the solution, and that data management software are key in optimizing for energy efficiency [17].

Reform. This aspect consists in reforming existing software components to minimize their energy use. They concern mainly two aspects: (1) the definition of cost models to predict energy and (2) the proposition of optimization techniques to reduce energy.

- **Energy cost models.** Prior works have been concentrated on building power cost models to predict query power consumption. In [47,48], the authors discussed the opportunities for energy-based query optimization, and a power cost model is developed in the conjunction of PostgreSQL's cost model to predict the query power consumption. A static power profile for each basic database operation in query processing is defined. The power cost of a plan can be calculated from the basic SQL operations, like CPU power cost to access tuple, power cost for reading/writing one page, and so on, via different access methods and join operations using regression techniques. The authors adapt their static model to dynamic workloads using a feedback control mechanism to periodically update model parameters using real-time energy measurements. The authors of [24] propose a technique for modeling the peak power of database operations. A pipeline-based model of query execution plans was developed to identify the sources of the peak power consumption for a query and to recommend plans with low peak power. For each of these pipelines, a mathematical function is applied, which takes as input the rates and sizes of the data flowing through the pipeline operators, and as output an estimation of the peak power consumption. The authors used piece-wise regression technique to build their cost model. In the same direction, the work of [25] proposes a framework for energy-aware database query processing. It augments query plans produced by traditional query optimizer with an energy consumption prediction for some specific database operators like select, project and join using linear regression technique. [35] attempts to model energy and peak power of simple selection queries on single relations using linear regression. In our previous works [37], we proposed cost models to predict the power consumption of single and concurrent queries. Our model is based on pipeline segmenting of the query and predicting their power based on its Inputs-outputs (IO) and CPU costs, using polynomial regression techniques.

- **Optimization techniques.** The presence of energy consumption cost models motivate the research community to propose cost-driven techniques. The work in [26] proposed an Improved Query Energy-Efficiency (QED) by *Introducing Explicit Delays mechanism*, which uses query aggregation to leverage common components of queries in a workload. The work of [25] showed that processing a query as fast as possible does not always turn out to be the most energy-efficient way to operate a DBMS. Based on their proposed framework, they choose query plans that reduce energy consumption. In [24], cost-based driven approach is proposed to generate query plans minimizing the peak power. In [38], genetic algorithm with a fitness function based on a energy consumption cost model, is given to select materialize views reducing energy and optimizing queries. The work by Xu *et al.* [47] is close in spirit to our proposal in this paper. They integrate their cost model into the DBMS to choose query plans with a low power at the optimization phase. However, they do not study the consumed energy at each phase of query optimizers. Moreover, they use a simple cost model that do not capture the relationship between the model parameters.

Evaluate. Energy evaluation is a sensitive point and as a consequence it requires accurate and *transparent* evaluation to show its savings. For transparent perspective, we propose an open platform available at the forge of our laboratory[6], allowing researchers, industrials and students to evaluate it.

---

[6] http://www.lias-lab.fr/forge/projects/ecoprod.

3 Background

To facilitate the understanding of our proposal, some concepts and definitions are given related to electrical energy and the life cycle of the data warehouse design. Electric current is the flow of electric charge measured in *amperes* (A). Amperes define the amount of electric charge transferred by a circuit per second.

Definition 1. *The energy is the total amount of work performed by a system over a period of time.*

Definition 2. *The power is the rate at which the system performs the work.*

Energy is usually measured in *joules* while power is measured in *watts*. Formally, energy and power can be defined as:

$$P = \frac{W}{T} \tag{1}$$

$$E = P \times T \tag{2}$$

Where P, T, W, and E represent respectively, a power, a period, the total work performed in that period of time, and the energy. Work is done at the rate of 1 W when 1 A is transferred through a potential difference of 1 V. [7].

The power consumption of a given system can be divided into two parts (i) baseline power and (ii) active power.

Definition 3. *The Baseline power is the power dissipation when the machine is idle. This includes the power consumption of the CPU, memory, I/O, fans, and other motherboard components in their idle state.*

Definition 4. *Active power is the power dissipation due to the execution of the workload. The active power component is determined by the kind of workload that executes on the machine and the way it utilizes CPU, memory, and I/O components.*

There exist two concepts of power that have to be considered during the evaluation of power utilization in DBMS. Average power representing the average power consumed during the query execution and peak power representing the maximum power. In this paper, we consider the average power.

The energy consumption can be reduced if either the average power consumption or the time intervals are reduced. Since optimizations in improving the performance of queries are widely studied, we focus on the power part of the Eq. 2 in this work. The first step is to model the power in order to estimate its consumption by the queries.

4 Eco-Query Optimizer Design

In order to build energy-aware query processing, we first propose an audit of their components to understand whether they may be energy-sensitive or not. After our audit, we present in details our framework for designing energy-aware query processing.

Recall that a query processing is responsible for executing queries respecting one or several non-functional requirements such as response time. The process of executing a given query passes through four main steps: (i) parsing, (ii) rewriting, (iii) planning and optimizing and (iv) executing (cf. Fig. 2). To illustrate these steps, we consider PostgreSQL DBMS as a case study.

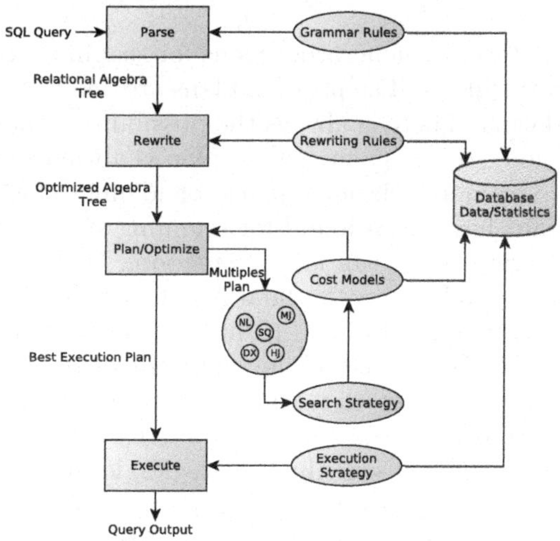

Fig. 2. Query optimizer steps

4.1 Parse

The parser has to check the query string for valid syntax using a set of grammar rules. If the syntax is correct, a *parse tree* is built up and handed back. After the parser completes, the transformation process takes the parse tree as input and does the semantic interpretation needed to understand which tables, functions, and operators are referenced by the query. The data structure that is built to represent this information is called the *query tree*. The cost of this phase is generally ignored by the DBMS since it finishes very quickly. We follow the same logic and suppose that the energy consumption is negligent.

4.2 Rewrite

The query rewrite processes the tree handed back by the parser stage and it rewrites the tree to an alternate using a set of rules. The rules are system or user defined. This rules-based phase is also used in materialization views query rewriting. As for the previous step, the cost is ignored due to the fast completion.

4.3 Plan/Optimize

The task of the planner/optimizer is to create an optimal execution plan. A given SQL query can be actually executed in different ways, each of which will produce the same set of results. The optimizer's task is to estimate the cost of executing each plan using a cost-based approach and find out which one is expected to run the fastest.

Plan. The planner starts by generating plans for scanning each individual relation (table) used in the query. The possible plans are determined by the available *indexes* on each relation. There is always the possibility of performing a *sequential scan* on a relation, so a sequential scan plan is always created. If the query requires joining two or more relations, plans for joining relations are considered after all feasible plans have been found for scanning single relations. The available join strategies are: *nested loop join*, *merge join*, *hash join*. When the query involves more than two relations, the final result must be built up by a tree of join steps, each with two inputs. The planner examines different possible join sequences to find the cheapest one. If the query uses less than a certain defined threshold, a near-exhaustive search is conducted to find the best join sequences; otherwise, a heuristics based genetic algorithm is used.

To study the effects of such searching strategies, let us consider the query *Q8* of the TPC-H benchmark[7]. This a complex query which involves the join of 7 tables. We modify the planner of PostgreSQL in three manners: (i) searching for a plan by employing actual DBMS strategy (ii) using the genetic algorithm, and (iii) manually by forcing the planner to choose a certain plan. For each strategy, we calculate the planning time, and the total energy consumption during query execution against 10 GB datasets. Results are presented in Table 1.

From the table, we can see that setting the query plan manually gives the betters results, in both time and energy. While the default searching algorithms (semi-exhaustive) leads to a slightly more execution time and energy consumption. The genetic algorithm gives the worst results in this example, perhaps due to the small number of tables in the query, since this strategy is used by the DBMS where there are more than 12 tables. Considering this small number of tables, if we go in real operational databases where there are a hundred of tables, the searching strategy used by the planner can lead to a noticeable energy

Table 1. Planning step for TPC-H *Q8* with different searching strategies.

Search algo	Planning time (s)	Energy (J)
Default	0.110006	5200.362
GA	0.977013	5387.648
Manual	0.092054	5160.036

[7] http://www.tpc.org/tpch/.

consumption. Thus, setting the query plan of queries manually by the database administrator is recommended in large databases to gain in energy efficiency.

Optimize. To evaluate the response time for each execution plan, cost functions are defined for each basic SQL operator. The general formula to estimate the cost of operator *op* can be expressed as:

$$Cost_{op} = \alpha \times I/O \oplus \beta \times CPU \oplus \gamma \times Net \tag{3}$$

Where I/O, CPU, and Net are the estimated pages numbers, tuples numbers, communication messages, respectively, required to execute *op*. They are usually calculated using database statistics and selectivity formulas. The coefficients α, β and γ are used to convert estimations to the desired unit (e.g., time, energy). $\oplus$ represents the relationship between the parameters (linear, non-linear). The coefficient parameters and their relationship can be obtained using various techniques such as calibration, regression, and statistics. Thus, an energy cost model must be defined at this stage with the relevant parameters.

The finished plan tree consists of sequential or index scans of the base relations, plus nested-loop, merge or hash join nodes as needed, plus any auxiliary steps, such as sort nodes or aggregate-function calculation nodes.

4.4 Executor

The executor takes the plan created by the planner/optimizer and recursively processes it to extract the required set of rows. This is essentially a demand-pull *pipeline* mechanism. Each time a plan node is called, it must deliver one more row, or report that it is done delivering rows. Complex queries can involve many levels of plan nodes, but the general approach is the same: each node computes and returns its next output row each time it is called. Each node is also responsible for applying any selection or projection expressions that were assigned to it by the planner[8].

To study the effect of the execution step on designing green-query processing, we consider an example of query *Q22* from the TPC-H benchmark. Figure 3 presents the execution plan returned by the PostgreSQL query optimizer. As we showed in [37], the power consumption is directly influenced by execution model of the DBMS. Therefore, the execution plan can be divided into a set of segments, we refer to these segments as *pipelines*, where the pipelines are the concurrent execution of a contiguous sequence of operators [16]. The pipeline segmentation of the optimizer plan for query *Q22* is shown in Fig. 3, there are 4 pipelines, and 4 terminal *blocking/semi-blocking* operations (`Aggregate`, `Hash`, `Sort` and `GroupAggragate`). For instance, the `Sort` is a blocking operator since it cannot produce any result until it reads and sorts all the input rows. The `Hash` is a semi-blocking operator, since it is blocking only in the build phase, it starts producing rows in the prob phase. On the other hand, a *non-blocking* operator

[8] https://www.postgresql.org/docs/current/static/executor.html.

consumes rows as input and produces rows as output at the same time (e.g.: `SeqScan`). Knowing this fact, a partial order of the execution of the pipelines is enforced by their blocking operators (e.g., PL3 cannot begin until PL2 is complete).

In our previous study, we showed that *when a query switches from one pipeline to another, its power consumption also changes.* During the execution of a pipeline, the power consumption usually tends to be approximately constant [37]. Therefore, the pipelining execution is very important and has a direct impact on power consumption during query execution. The design of green query processing should take into consideration the execution strategy, which is unfortunately ignored by Xu *et al.* [47].

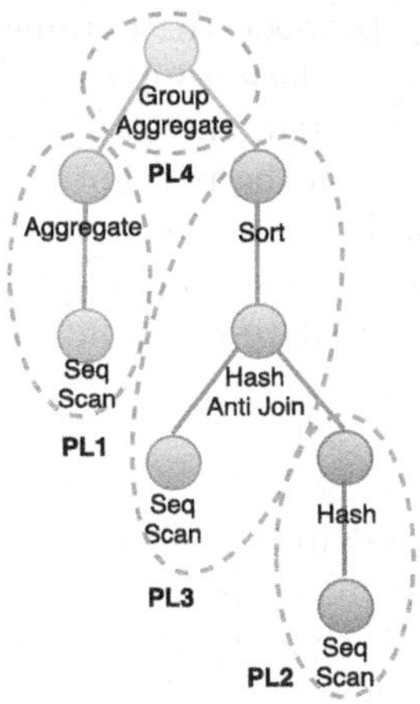

Fig. 3. Execution plan of TPC-H benchmark query *Q22* with corresponding pipeline annotation.

4.5 Our Energy-Aware Query Processing

In this section, we describe the design and the implementation of our proposal into PostgreSQL database. As we mentioned above, the planner/optimizer and the executor stages have an impact on energy consumption and should be considered in designing any green-query processing. We extended the cost model, the query optimizer and the communication interface of PostgreSQL to include the energy dimension.

Inspired by the observation made in the previous section, we designed our cost-based power model. The basic idea of this model is to decompose an execution plan into a set of power independent pipelines delimited by blocking/semi-blocking operators. Then for each pipeline, we estimate its power consumption based on its CPU and I/O cost.

The work-flow of our methodology is described in Fig. 4.

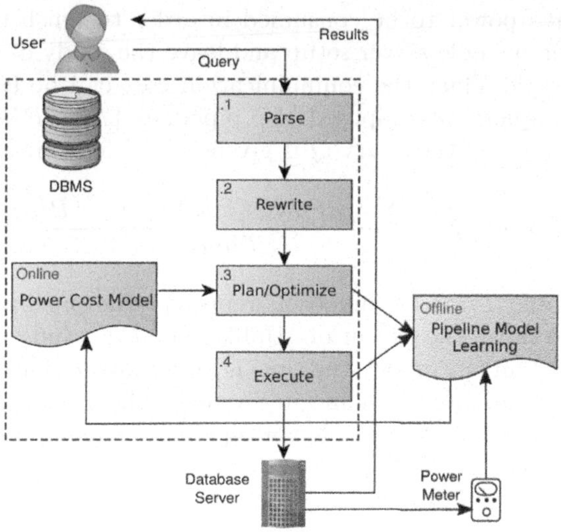

Fig. 4. The design methodology

4.6 Power Cost Model

In this section, we present our methodology for estimating energy consumption. The characteristics of our model include: (i) the segmentation of an execution plan into a set of pipelines, (ii) the utilization of the pipeline cost to build the regression model, and (iii) the estimation of the power of future pipeline based on pipeline cost and the regression equation.

Pipeline Segmentation. When a query is submitted to the DBMS, the query optimizer chooses an execution plan (cf. Fig. 3). A physical operator can be either *blocking* or *nonblocking*. An operator is blocking if it cannot produce any output tuple without reading at least one of its inputs (e.g., sort operator). Based on the notion of blocking/nonblocking operators, we decompose a plan in a set of pipelines delimited by blocking operators. Thus, a pipeline consists of a set of concurrently running operators [13]. As in previous work [13], the pipelines are created in an inductive manner, starting from the leaf operators of the plan. Whenever we encounter a blocking operator, the current pipeline ends, and a new pipeline starts. As a result, the original execution plan can be viewed as a tree of pipelines, as showed in Fig. 3.

Model Parameters. Given a certain query, the query optimizer is responsible for estimating CPU and I/O costs. Our strategy for pipeline modeling is to extend the cost models that are built into the PostgreSQL DBMS for query optimization. To process a query, each operator in a pipeline needs to perform CPU and/or I/O tasks. The cost of these tasks represents the "cost of the pipeline", which is the active power to be consumed in order to finish the tasks. In this paper, we focus on a single server setup and leave the study of distributed databases as future work. Thus, the communication cost can be ignored. More formally, for a given query Q composed of p pipelines $\{PL_1, PL_2, \ldots, PL_p\}$. The power cost $Power(Q)$ of the query Q is given by the following equation:

$$Power(Q) = \frac{\sum_{i=1}^{p} Power(PL_i) * Time(PL_i)}{Time(Q)} \tag{4}$$

The $Time(PL_i)$ and $Time(Q)$ functions represent the pipelines and the query estimated time to finish the execution. Unlike Xu *et al.* study which ignores the execution time [48], in our model, the time is an important factor in determining the CPU or I/O dominated pipeline in a query. The DBMS statistics module provides us with this information. Let a pipeline PL composed of n algebraic operations $\{OP_1, OP_2, \ldots, OP_n\}$. The power cost $Power(PL)$ of the pipeline PL is the sum of CPU and I/O costs of all its operators:

$$Power(PL) = \beta_{cpu} \times \sum_{i=1}^{n} CPU\_COST_i + \beta_{io} \times \sum_{i=1}^{n} IO\_COST_i \tag{5}$$

where β_{cpu} and β_{io} are the model parameters (i.e., unit power costs) for the pipelines. For a given query, the optimizer uses the query plan, cardinality estimates, and cost equations for the operators in the plan to generate counts for various types of I/O and CPU operations. It then converts these counts to time by using system-specific parameters such as CPU speed and I/O transfer speed. Therefore, in our model, we take I/O and CPU estimations already available in PostgreSQL before converting it to time. The IO_COST is the predicted number of *I/O* it will require for DBMS to run the specified operator. The CPU_COST is the predicted number of *Tuples* it will require for DBMS to run the specified operator. A summary of the formulas used to calculate I/O and CPU costs for each basic operator can be found in Table 2b with the symbols listed in Table 2a. For instance, if the operator i is a sequential scan, its I/O cost equals to p_{seq}, and CPU cost equals to t_{seq}. These two functions are the same used by PostgreSQL query optimizer to estimate the cardinality. The parameters are based on the database statistics module, and on our proposed models.

Parameters Calibration. The key challenge in Eq. (5) is to find model parameters β_{cpu} and β_{io}. Simple linear regression technique, as used in [24,25,48], did not work well in our experiments, especially when data size changes, this is because the relationships between data size and power are not linear. In other words, processing large files does not *always* translate in high power consumption. It depends more on the type of queries (I/O or CPU intensive) and their execution time. Therefore, we employed multiple polynomial regression techniques. This method is suitable when there is a *nonlinear* relationship between the independents variables and the corresponding dependent variable. Based on

Table 2. Cost model calculation formulas and parameters.

Parameter	Definition	
(a) Cost model parameters notation		
β_{cpu}	power to perform one I/O op	
β_{io}	power to perform one CPU op	
m	buffer memory size	
$block$	DBMS page size	
T_i	the size of table i	
t_i	# of input tuple for the op i	
p_i	# of input pages for the op i	
f	index selectivity	
s	input relation size for the sort op	
n_{hash}	# of clauses in building phase	
p_{hash}	# of partitions in probing phase	
$p_{outer/inner}$	# of pages retrieved for join op	
$t_{outer/inner}$	# of tuples retrieved for join op	
n_{group}	# of grouping columns	
Parameter	I/O Cost	CPU Cost
(b) Cost model parameters for SQL operators		
Sequential scan	p_{seq}	t_{seq}
Index scan	p_{index}	$t_{index} \cdot f$
Bitmap scan	p_{bitmap}	$t_{bitmap} \cdot f$
Nested loop join	$p_{outer} + p_{inner}$	$t_{outer} \cdot t_{inner}$
Sort merge join	$p_{outer} + p_{inner}$	$t_{sort(outer)} + t_{sort(inner)}$
Hash join	$p_{outer} + p_{inner}$	$t_{outer} \cdot n_{hash} + t_{inner} \cdot p_{hash}$
Sort	$p_{sort}, s < m; 0, else$	$t_{sort} \cdot log_2(t_{sort})$
Aggregate	0	$t_{agg} \cdot$
Group by	0	$t_{group} \cdot n_{group}$

our experiments, the order $m = 4$ gives us the best results (the residual sum of squares is the smallest). The power cost $Power(PL)$ of the pipeline PL is computed as:

$$\begin{aligned} Power(PL) = \beta_0 + \beta_1(IO\_COST) + \beta_2(CPU\_COST) + \\ \beta_3(IO\_COST^2) + \beta_4(CPU\_COST^2) + \\ \beta_5(IO\_COST \times CPU\_COST) + \\ \cdots + \\ \beta_{13}(IO\_COST^4) + \beta_{14}(CPU\_COST^4) + \epsilon \end{aligned} \tag{6}$$

Where IO_COST, CPU_COST denote the pipeline I/O and CPU costs respectively, these costs are calculated using the DBMS cost model functions, and ϵ is a noise term that can account for measurement error. The β parameters are regression coefficients that will be estimated while learning the model from training data. Thus, the regression models are solved by estimating the model parameters β, and this is typically done by finding the least-squares solution [29].

4.7 Plans Evaluation

The query optimizer evaluates each possible execution path and takes the fastest. Adding the energy criterion, we must adjust the comparison functions to reflect the tradeoffs between energy cost and processing time. The goal of our new multi-objectives query optimization problem (MOQP) is to select query plans that minimize both execution time and power consumption. The traditional performance cost model $Time(Q)$ for a query Q composed of n algebraic operations $\{OP_1, OP_2, \ldots, OP_n\}$, is defined as follows:

$$Time(Q) = \alpha_{cpu} \times \sum_{i=1}^{n} CPU\_COST_n + \alpha_{io} \times \sum_{i=1}^{m} IO\_COST_i \tag{7}$$

Where IO_COST, CPU_COST are already described. The α parameters are coefficients specified to our test machine, used to convert the query costs to time values. α_{cpu} is the *CPU* time required to execute one *CPU Cycle* and α_{io} is the *IO* time needed for the device to execute one *I/O* operation.

In order to solve the MOQP, we propose to use the weighted sum of the cost functions method to give the database administrator a solution with the desired trade-off. In this scalarization method, we calculate the weighted sum of the cost functions (i.e. Eqs. (5) and (7)) so as to aggregate criterions and have an equivalent single criterion to be minimized. In our case, this method can be defined as follows:

$$\begin{aligned} \text{minimize} \quad & y = \omega_1 \times Time(Q) + \omega_2 \times Power(Q) \\ \text{such} \quad & \omega_1 + \omega_2 = 1 \end{aligned} \tag{8}$$

Where ω_i are the weighting coefficients representing the relative importance of the time and power cost functions respectively. y represents the cost of query plan Q. We implemented these two coefficients as an external parameter in the DBMS, so the database administrator or users can change them on the fly.

Figure 5 shows the optimal query plan returned by the modified query planner/optimizer for TPC-H query *Q3* and how it changes when user preferences vary. Initially, we used a performance only optimization goal, the total estimated processing cost is 371080 and the total estimated power is 153. Changing the goal to be only power, the processing cost increased to 626035 but the power falls down to 120. In the trade-off configuration, the processing cost is 377426 and the power is 134. In Fig. 5(a), the nested loop operator draws the high amount of power in the query (33 watts) but the plan is chosen by the optimizer because

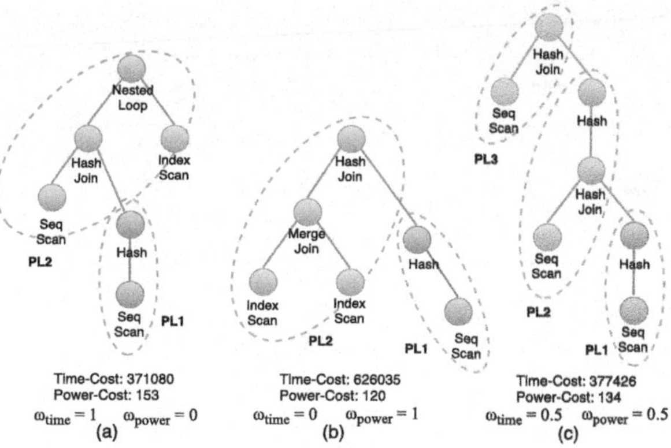

Fig. 5. The optimal plan for TPC-H query *Q3* when changing user preferences.

it is very fast. In Fig. 5(b), we realize that the merge join operator is the slowest in the query, its processing cost is 539200, with its power being minimal. The two hash join operators used in Fig. 5(c) give a good trade-off, for a 1.7% of performance degradation, we get 12.4% of power saving.

4.8 EnerQuery GUI

In this section, we describe the graphical user interface part of EnerQuery. The GUI helps to manipulate EnerQuery, changing parameters and showing in real time their impact on the power consumption.

The EnerQuery GUI interface is used to facilitate users manipulating the framework settings and seeing their effect on the system. The interface is implemented using C++ programming language and Qt library. Figure 6 gives an overview of the main GUI, which comprises several component modules:

Configuration. This module is responsible for the connection establishment with the DBMS server. Users can also specify the path for the power meter driver in order to capture real-time power consumption. The most important part here is the power/performance settings, which decide the optimization goals to be performance or power oriented.

SQL Query. In this module, users can give their SQL query to be executed. Queries supported vary from simple transactional operations to very complex reporting operations involving many tables with large data size. The execution is done in a separate thread and the results are displayed in a tree table widget.

Power Time-Line. When the user executes a query, EnerQuery dynamically displays via the power meter the real-time power consumption. After the query finishes executing, the total energy that has been consumed during query execution time is computed and shown. This can give users a real observation of the

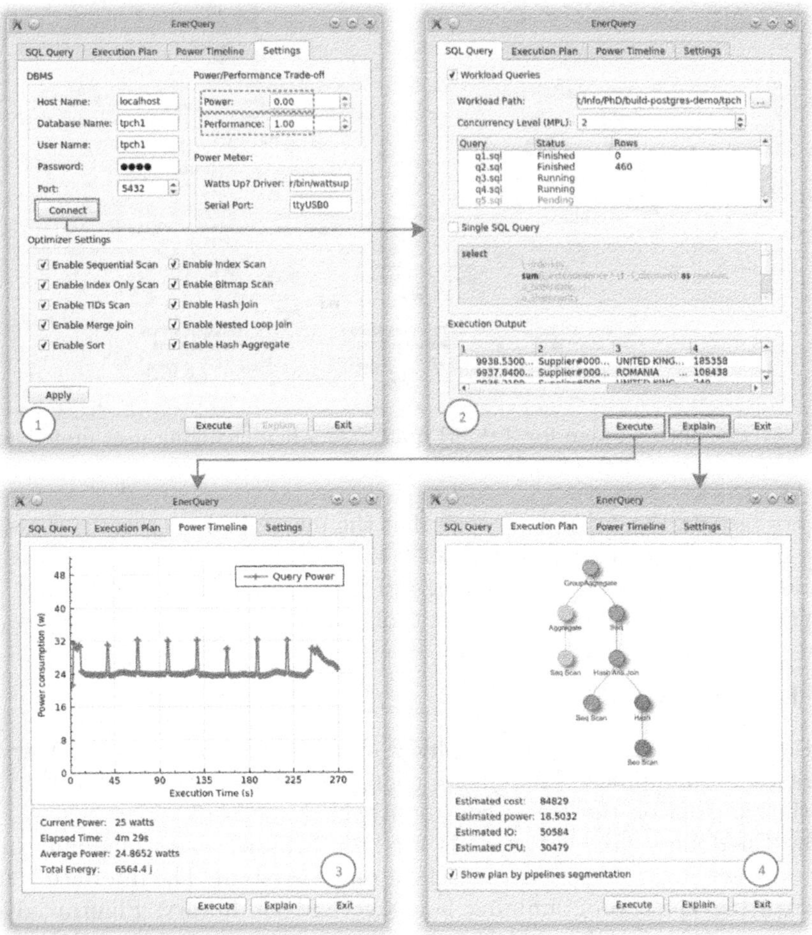

Fig. 6. EnerQuery main GUI and its component module panels.

energy that has been saved using the desired trade-off parameters. Also, users can compare the estimated and the real values to check the model accuracy or further refine it.

Execution Plan. When the user submits a query, the query optimizer will select their best execution plan with respect to the pre-defined trade-off. The execution plan is displayed with various information, such as estimated cost, power consumption, I/O and CPU costs for every physical operator through mouse-hovering events. Also, the pipeline notation is demonstrated, as shown in Fig. 6(4), the pipeline trees are grouped with the same color. The GUI shows how the trade-off parameters affect the generated plan. Thus, we can help users better understand and interpret runtime optimization information's and pipeline notation.

5 Energy Incorporation in Logical and Physical Phases

In the previous sections, we focused on the incorporation of the energy in query optimizers as done by some important works on energy on the context of databases. In this section, we propose to go further and considering energy during the design of a $\mathcal{DW}$. We focus on two main phases which are the logical and physical design.

5.1 Logical Design

There are three main approaches to represent logical model of a $\mathcal{DW}$, depending on how the data cube is stored:

- *(i) the relational ROLAP* (ROLAP), which stores data cube in relational databases and uses an extension of SQL language to process these data. Two main schemes are offered by ROLAP: the star schema and snowflake schema. A star schema consists of a one or several large fact table (s) connected to multiple dimension tables via foreign keys. Dimension tables are relatively small compared to the fact table and rarely updated. They are typically non normalized so that the number of needed join operations are reduced. To avoids redundancy, the dimension tables of a star schema can be normalized. There is debate on the benefits of having such normalized dimension tables, since it will, in general, slow down query processing, but in some cases it provides a necessary logical separation of data such as in the case of the demographic information [27].
- *(ii) multidimensional OLAP (MOLAP)* stores cube data in multidimensional array format. The OLAP operations are easily implemented on this structure. For high dimensionality of data, ROLAP solutions are recommended [45]. (iii) *Hybrid OLAP (HOLAP)* combines both approaches. It gets benefit from the storage capacity of ROLAP and the processing power of MOLAP.

The $\mathcal{DW}$ is well known by the hierarchies and dependencies between properties [45].

Figure 7 provides an overview of the global schema covering the domain of the Star Schema Benchmark $\mathcal{SSB}$, which is used further down for our experiments, and right below to illustrate the different types of existing correlations that we have identified and classified that way [9]:

- *Definition/equivalence relations or Generalization (DEF)*: when concepts/ roles are defined in terms of other concepts/roles. E.g. a *Supplier* is a *TradeAgent* that *Supplies* a *LineItem*.
- *Inclusion dependencies or Specialization (ID)* also called is-a relation or subsumption: it occurs when a concept/role is subsumed by another concept/role. E.g. *Customer* subsumes *TradeAgent*. When it concerns attributes, there is another application of this type: the notion of foreign keys, which states that the domain of one attribute must be a subset of the other correlated attribute.

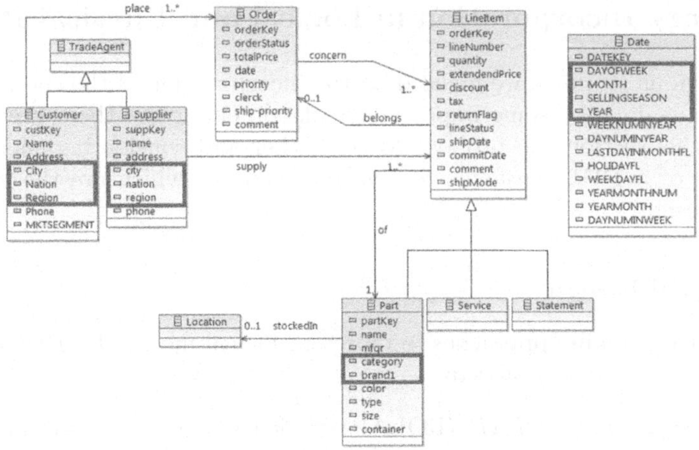

Fig. 7. *SSB* ontology

- *Functional dependencies (CD/FD)*: we note *CD* if concepts and *FD* if attributes. They figure when a set of concepts/roles (or their instances) *determine* another set of the same type. E.g. *custKey* determines the *name* of *Customer*.
- *Multi-valued dependencies (MD)* or *soft dependencies*: specific to attributes, it is a generalization of the functional dependencies. Formally, the difference between the two is the abolition of the determination criterion, in other words, to a value set, we can associate more than one value set of the same type.
- *Conditional Integrity constraints (CIC)*: specific to attributes, they denote the $\mathcal{DB}$ integrity constraints (algebraic or semantic)[9] involving more than one attribute [34] and holding on instances of the relations. Note that definitions and dependencies are considered as simple integrity constraints which are valid on entire relations, contrary to conditional ones where the correlation is accompanied with condition(s) that must be respected. This latter aspect moves the application level from attribute range level to attribute values level. In other words, only a subset of the member attributes domain is involved (reduced range). E.g. $Customer.City = Paris \rightarrow LineItem.discount \succ 20\%$. We distinguish two main categories: (i) conditional functional dependencies (*CFD*) [8] whereby the (*FD*) has to hold on a subset of the relation (a set of tuples) that satisfies a specific attribute pattern ([*Customer.city* = *London*, *Customer.name*] → [*Customer.phone*]), rather than on the entire relation *Customer*, and (ii) more specifically, association rules that apply for particular values of some attributes [2] (*Part.color* = 'red' → *Part.size* = 50).
- *Hierarchies (H)*: specific to attributes, and more present in $\mathcal{DW}$s, where a set of attributes makes up a dimension hierarchy (e.g. Industry, category, product). They can be assimilated to the *part-whole* relationships.

[9] IC specify conditions/propositions that must be maintained as true ($Part.size > 0$).

The particularity of this type, is that we could plan the construction of a new class for each hierarchy level.

From the former classification, we can infer the results achieved by exploiting these correlations throughout the design life-cycle of $\mathcal{DB}$. Those results belong to either conceptual, logical or physical levels. In more details:

- Conceptual level: the correlations having impact on the definition of conceptual schema are of type: DEF or ID. This impact consists of creating new concepts/roles (non canonical) when using DEF, or creating subsumption relationships linking the concepts/roles when using ID.
- Logical level: exploiting correlations of type CD, FD or H, has a direct impact on logical level: data normalization when using FD or CD, multidimensional $\mathcal{OLAP}$ annotation, hierarchical $\mathcal{OLAP}$ normalization when using H.
- Physical level: a lot of studies have exploited correlations of type MD, ID or CIC in the definition of the Physical Design Structures ($\mathcal{PDS}$).

In the light of the foregoing, we believe that any evolution/transition throughout the design life-cycle of $\mathcal{DB}$ can be controlled by correlations. Table 3 shows different studies in this field. In fact, thanks to the formal power of ontologies, and their strong similarity with conceptual models, we can store correlations (identified by the $\mathcal{DB}$ users notably the designer) right from the conceptual phase. Afterwards, the transition to the logical level is henceforth based on correlations: namely the dependencies (CD, FD) for $\mathcal{DB}$, and hierarchies for $\mathcal{DW}$, as for the transition to the physical, it becomes controlled by either MD, ID or CIC. Indeed, several studies have shown that $\mathcal{DB}$ performance can be vastly improved by using $\mathcal{PDS}$ defined upon correlations, and even more when exploiting the interaction - generated upon correlations - between these $\mathcal{PDS}$, as is the case concerning $\mathcal{MV}$ and indexes in the *CORADD (for CORrelation Aware Database Designer for Materialized Views & Indexes) system* [22] (see Table 3).

Table 3. Related work on correlations exploitation over the design life-cycle of $\mathcal{DW}$

Studies\Phases	MC	ML	MP	OLAP	Other
Anderlik et al. [3]				DEF/ID Roll-up	
Stohr et al. [43]		H Fragmentation			
Kimura et al. [22]			FD/MD $\mathcal{MV}$/indexes		
Brown et al. [12]			CIC Query optimizer		
Agrawal et al. [2]					CIC Data-mining
Petit et al. [31]	ID/FD ER schema	ID/FD Relational schema			ID/FD Reverse engineering

5.2 Physical Design

It is a crucial phase of the $\mathcal{DW}$ life cycle. Note that the majority of non-functional requirements are evaluated during this phase. It uses the inputs of deployment and the logical phases. In this design, optimization structures such as materialized views, indexes, data partitioning, etc. are selected to optimize one or several non-functional requirements such as query performance and energy.

5.3 Summary

Based on the above discussion, we figure out that the energy may be included in all phases, and especial logical and physical ones. Recall that the physical phase in the funnel of all phases. This is characterized by the entries of its corresponding problem, called the physical design problem ($\mathcal{PDP}$). These entries are:

1. the chosen logical model ($\mathcal{LM}$) of the $\mathcal{DW}$ application identified during the logical phase of the life cycle;
2. the used query language ($\mathcal{QL}$) offered by the DBMS;
3. the used deployment platform ($\mathcal{DP}$) including the target DBMS and hardware;
4. the available logical and physical optimizations ($\mathcal{PO}$) offered by the DBMS that have to be exploited by DBA either in isolated or joint manners to ensure the performance of her/his workload. The logical optimizations refer to those offered by query optimizers (see Sect. 4). Whereas, the physical optimizations refer to the optimization structures ($\mathcal{OS}$) (e.g. materialized views, indexes, etc.) selected during the physical design phase.

The formalization of the $\mathcal{PDP}$ by taking into account the two non-functional requirements representing the query processing and the energy consumption is as follows: Given:

- a *workload* $\mathcal{W}$ of numerous queries, expressed in:
- a *query language* ($\mathcal{QL}$) related to:
- a *logical model* ($\mathcal{LM}$) translated to:
- a *physical model* ($\mathcal{PM}$) associated to:
- a set of *physical optimizations* ($\mathcal{PO}$) deployed in:
- a *platform* ($\mathcal{P}$).

$\mathcal{PDP}$ aims at picking optimization structures satisfying both requirements and respecting the given constraints (e.g., the storage). This is known to be NP-Complete [32]. The resolution of our problem has to *exploit as much as possible the dependencies between inputs and their variabilities.* Figure 8 presents an UML model illustrating our proposal.

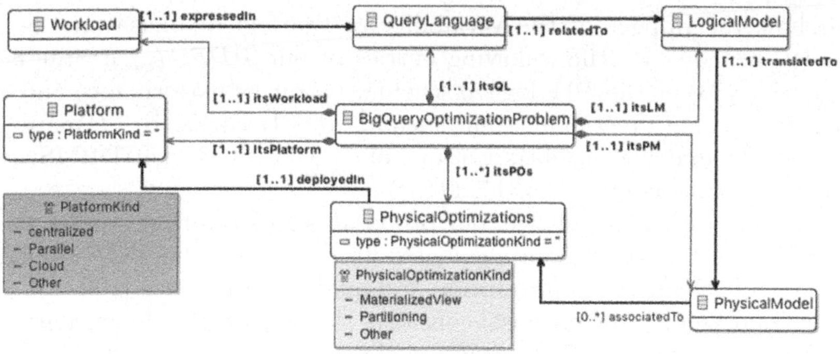

Fig. 8. An UML model of our big-queries optimization problem

5.4 Energy at Logical and Physical Phases

To integrate the energy when logical designing a $\mathcal{DW}$, we can exploit the relationships that may exist between its properties. An analogy can be immediately drawn between the theory of the variability management ($\mathcal{VM}$) [4] (which is widely studied by the community of Software Engineering) and $\mathcal{PDP}$. In fact, this latter owns different variation points (dimensions, entries), as depicted in Fig. 9. Each variation point is seen as a complex search problem, often using a mathematical non-functional requirement-driven cost model to evaluate and select "best" solutions.

Based on this figure, we can easily identify the *dependencies* among these dimensions. Actually, varying the logical schema strongly impacts the following entries: *workload*, the optimization structures ($\mathcal{OS}$) and the *constrains.*

Nevertheless, designers still intuitively fix a logical solution out of the large panoply of schemes, hence omitting eventual more relevant alternatives ($\mathcal{VM}$ aspect). Bearing this in mind, we fix some objectives to handle this missing piece in $\mathcal{PDP}$ puzzle: **(i)** capturing variability, **(ii)** *studying the impact* of variability on dependent components, as well as *efforts in terms of modeling and coding* to be spent by designers to manage this variability, and **(iii)** validating our methodology.

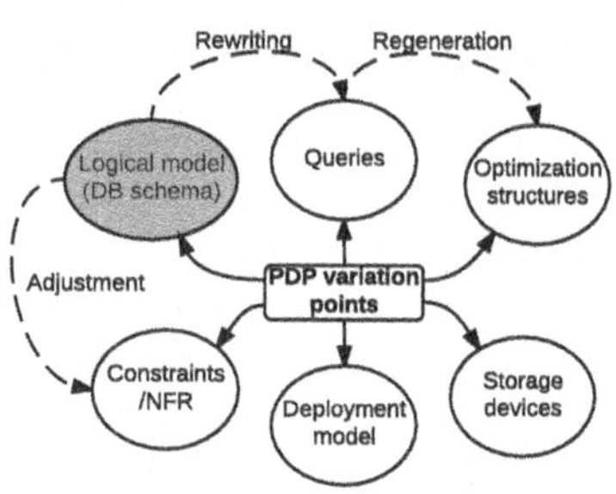

Fig. 9. Variation points of physical design problem.

To show the impact of the variability of the logical phase on the physical model, we consider the following entries of our $\mathcal{PDP}$: *(a)* a Star Schema Benchmark (SSB) as the $\mathcal{DW}$ logical schema, *(b)* query performance and energy consumption as non-functional requirements, *(c)* SSB workload, *(d)* materialized views as a $\mathcal{OS}$ and logical optimizations offered by the target DBMS, *(e)* the storage constraint dedicated to this $\mathcal{OS}$.

To the best of our knowledge, our proposal is the sole that studies the variation of the logical schema and evaluates its impact on the $\mathcal{PDP}$ dimensions, according to both energy consumption and query performance [10].

As we said before, our proposal considers the majority of dimensions of variability axes. To show the real impact on the physical design and to reduce the complexity of treating all dimensions at once, we incrementally integrate dimensions. We start by evaluating the impact of the logical schema variation on the physical design when executing a workload, without considering optimization structures. Only optimizations offered by target DBMS (referred to as *logical optimization*) are considered. Secondly, we evaluate the impact of the logical variation on the problem of materialized view selection.

Before evaluating the impact of the logical schema variation, we first have to explain the origin of this variation.

5.5 Capturing Variability of Logical Design

There is a patent lack of considering logical design variability in $\mathcal{PDP}$. Yet while most problems can be solved by fine-tuning the physical schema, some performance/NFR problems are caused by a non-optimized logical schema. In fact, this latter is variable (can have many variants) thanks to the so-called *correlations*: Functional dependencies, Multi-valued dependencies and hierarchies (cf. Sect. 5.1). In case of data-warehouses for instance, let $\mathcal{LM} = \{F, D_1, D_2, ..., D_n\}$ be the DW logical schema from which different other schemes (star and snowflake) can be generated. F for fact table, D_i for dimensions. The search space (referred to as product line scope in variability jargon) can be estimated around $\prod_{\{d=1\}}^{n} 2^{h_d - 1}$ possible $\mathcal{LM}$s, such that h_d is the number of hierarchical levels of a dimension $d/d \in \{1..n\}$. The generation process (step 1 of Algorithm 1) is based on attributes correlations [9].

5.6 Scenario 1: Impact of $\mathcal{VM}$ on Logical Optimizations

To study the impact of variability of the logical model on the physical phase, we consider a *naive scenario*, where logical optimizations (e.g. join implementations, join ordering, etc.) are delegated to the query optimizer of the target DBMS (Oracle 11gR2 in our case) and advanced optimizations structures such as materialized views are absent. We consider the query performance and energy consumption as two *objective functions* when executing the workload. In practice, for each variant of the initial logical schema of our $\mathcal{DW}$, we compute both metrics. Note that each variant requires rewriting efforts of the initial workload. Algorithm 1 gives an overview of our approach:

Algorithm 1. Algorithm dedicated to Scenario 1.

Input: $\mathcal{DW}$ logical model: $\mathcal{DW} = \{F, D_1, D_2, ..., D_n\}$; $\mathcal{Q} = \{q_1, q_2, ..., q_m\}$;
Output: $\mathcal{DW}\prime$: $\mathcal{DW}$ logical schema having the *most suitable* performance/power-saving trade-off
Generate the different possible logical schemes;
for *each generated schema* **do**
  Calculate the size of the schema;
  for *each query in the workload* **do**
    Rewrite the query conforming to the target schema;
    Execute the query;
    Record the overall query power & its execution time;
  Calculate the time and power averages of queries;
Normalize power and time values;
Weight both objectives (power & time);
$\mathcal{DW}\prime$ = Schema having the minimum of the weighted sum;

Our algorithm provides us both metrics corresponding to query performance and energy consumption. In order to help designers choose the schema that best fits their requirements, we initially propose to use the weighted sum of the objective functions method that allows formulating the desired trade-off between target non functional requirements (Eq. (8)). For example, an eco-performing schema would have an $\omega_{pow} = \omega_{perf} = 0.5$) while a performance-oriented schema would have $\omega_{perf} > \omega_{pow}$, contrary to an eco-oriented schema ($\omega_{pow} > \omega_{perf}$). This technique is well suited when the *Pareto* front is convex, which is the case with our curve, as further illustrated in Sect. 5.8.

5.7 Scenario 2: Impact of $\mathcal{VM}$ on Physical Optimizations

In this scenario, we leverage the previous one by considering an optimization structure representing materialized views [28]. In our study, we do not delegate the selection of materialized views to advisors of commercial DBMSs, we propose instead an algorithm selecting them according to our previous metrics. This selection is proven to be NP-hard problem, and has been subject to many studies [28]. The process of selecting views requires three main components [11]:

(a) A data structure to capture the interaction among queries, like the *And-Or view graph* or *Multi-View Processing Plan* [28]. It puts the algebraic operations of queries all together in a certain order as an acyclic graph. Starting from base tables as leaf nodes to queries results as root nodes, through intermediate nodes: unary operations (like selection/projection) and binary ones (like join/union). Getting the optimal order between intermediate nodes – join ones in particular- determines the efficiency of the structure.
(b) Algorithms (e.g. deterministic algorithms, randomized algorithms, etc. [11]) exploiting a such structure to pick the best configuration of materialized views.
(c) Cost models estimating different non functional requirements.

(a) The construction of the data structure. Our proposal has the ability to consider very large number of queries. This is due to the data structure borrowed from hyper-graphs [11] representing the global plan of the workload as well as the interaction among queries. The main steps of the process of selecting views are [11]:

- **Step 1:** Parse query workload to extract the different algebraic operations;
- **Step 2:** Construct the hypergraph He out of join nodes, such that every He represents a query and thus contains its different join nodes modeled as vertices;
- **Step 3:** Partition He into a set of connected components He_{sub} (disjoint sub-hypergraphs), to put interacting nodes together;
- **Step 4:** Now that the workload is modeled by a partitioned hypergraph, an order between join nodes in each sub-hypergraph has to be established. By definition of MV, the join node(s) that maximize the reuse benefit have to be materialized. A *benefit function* is used in order to calculate the benefit of reusing every node and hence to order them;
- **Step 5:** Merge the resulting sub-hypergraphs He_{sub} to generate the global structure.

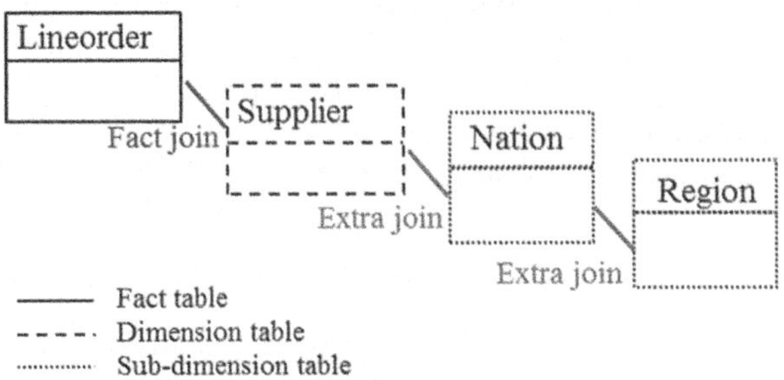

Fig. 10. Explanation of join types.

Our contribution concerns the He construction, and the ordering of He_{sub} nodes (2nd & 4th steps). In fact, only star schemes are handled by the baseline approach [11], unlike ours that considers any multidimensional model (star, snowflake, constellation). The difference lies in that there henceforth exists some extra-join nodes not involving any-more the fact table (joins between dimensions and sub-dimensions, which we have called *extra-joins* against *fact-joins* as shown in Fig. 10), and this leads to a totally different situation. As depicted in Fig. 11, the 2nd and the 4th arrangements are impossible configurations in the baseline approach and frequent ones in ours. Indeed, we will have more than one starting nodes in one connected component.

However, the extra-joins introduce a partial order since they have to precede the fact-ones (sub-dimensions must be joined before their dimensions). This

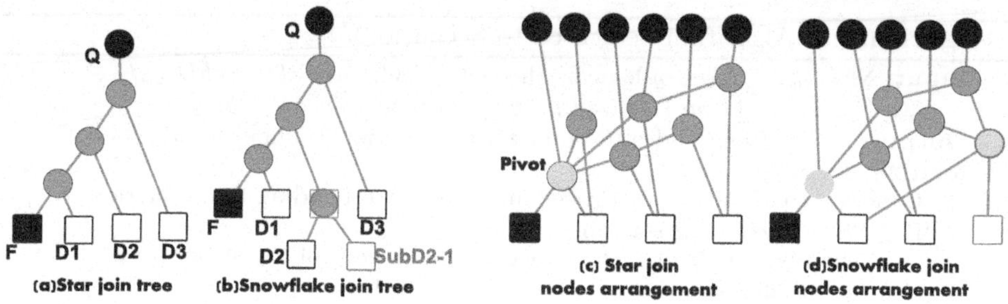

Fig. 11. Possible join nodes arrangement in a global plan.

order must be considered when ordering nodes of He_{sub} (step4) so that: (i) the mother dimension (in Fig. 10, *Supplier* is the mother dimension of *Nation* and *Region*, since these latter are derived from it) must always appear after its extra-joins, (ii) the outer extra joins must always figure before the inner ones. These rules are guaranteed thanks to the below benefit functions, the purpose of which, is to find the pivot nodes and thus order the He_{sub} nodes.

Given: n_{f_i} a fact-join, n_{e_i} an extra-join, k the number of extra joins implied by a fact-join, *nbr* the number of queries using the node in question, *cost* its processing cost $\Rightarrow$

$$\begin{cases} cost_{total}(n_{f_i}) = cost(n_{f_i}) + \sum_{j=1}^{k} cost(n_{e_j}) \\ benefit(n_{e_i}) = (nbr - 1) * cost(n_{e_i}) - cost(n_{e_i}) \\ benefit(n_{f_i}) = (nbr - 1) * cost(n_{f_i}) + \\ \qquad \sum_{j=1}^{k}(cost(nbr - 1) * cost(n_{e_j})) \\ \qquad -cost(n_{f_i}) - \sum_{j=1}^{k} cost(n_{e_j}) \end{cases}$$

(b) Materializing nodes and schema selection (Algorithm). Our approach, as summarized in Algorithm 2 ($\mathcal{LM}$ and $\mathcal{Q}$ are generated from Algorithm 1), is based on the *hyper-graph based* structure. In fact, if designer looks primarily for optimizing query performance, we will create this *structure* for each schema among the *top-k* performance-oriented schemes, materialize the pivot node of each one (the most advantageous node), execute queries for each schema and finally compare results. The schema having the smallest execution time of its queries will be the selected one. Otherwise, if designer needs to optimize both query performance and energy saving, the *structures* of the *top-k* eco-oriented schemes (or trade-off-oriented according to designer needs) will be generated.

Materializing the pivot node does not make sense anymore for saving power, because the pivot node for MV is the node having the best reuse benefit which does not necessarily give the optimal power values. Materializing all the join nodes is neither the solution because this would entail the highest power consumption [38]. Testing 2^n possible configurations, where n is the number of join nodes, to find *Pareto* solutions, is impossible especially in DW workloads involving a lot of joins. A Pareto solution is a set of nodes (a view configuration), that

Algorithm 2. Algorithm dedicated to Scenario 2.

Input: NFR, $\mathcal{LM}$: a set of logical schemes / $\mathcal{LM}_i = \{F, D_j, SubD_{jk}\}/j \in \{1..n\}, k \in \{1..n_k\}, \mathcal{Q}_i = \{q_{i_1}, q_{i_2}, ..., q_{i_m}\}$
Output: A set of join nodes to be materialized (view configuration)
for $\mathcal{LM}_i \in \mathcal{LM}$ **do**
  Generate the Multi View Processing Plan corresponding to its queries $\mathcal{Q}_i$;
  if *NFR = performance* **then**
    Each pivot node of each connected component of the structure is materialized;
  else /* NFR = *energy & performance* */
    Annotate each join node by its execution time and power consumption;
    Apply an evolutionary algorithm to select candidate views to be materialized optimizing *performance* as well as *energy*;
    Apply the weigthed sum on these candidates to select one view configuration;

would give - when materialized - values that can not be improved without making at least power or performance worse off. They are the best alternative since there does not typically exist a solution that minimizes all objective functions at once. Evolutionary Algorithms (*EAs*) are indeed suitable for multi-objectives optimization problems where large search spaces can be handled and multiple alternative trade-offs can be generated in a single optimization run [50]. The general idea behind an *EA* is to investigate a set of solutions that represent the Pareto optimal set as well as possible.

(c) Energy cost-model. To evaluate the interest of the presence of a materialized view, without deploying, each time, the $\mathcal{DW}$ (schema and optimization structures), we adjust our mathematical cost model developed in [38]. In fact, this cost model has been constructed by assuming a DW with star schema. As a consequence, our adaptation consists in making it more generic to consider all variants of the logical schemes. This adaption mainly concerns the training phase that allows identifying the relevant parameters of our cost models using *polynomial* multivariate regression model [38].

5.8 Experimental Study

To evaluate the logical variability impact on physical design, we conduct intensive experiments related to our two scenarios. First, we present our development environment including hardware, software, datasets, and results.

Hardware Setup. We used a Dell PowerEdge R210 II workstation having an Intel Xeon E3-1230 V2 3.30 GHz processor, 10 GB of DDR3 memory and a 2 × 500 GB hard drive. Our machine is equipped with a "Watts UP? Pro ES[10]" power meter

[10] https://www.wattsupmeters.com/.

with one second as a maximum resolution. As commonly set up, the device is directly placed between the power supply and the $\mathcal{DB}$ workstation under test to measure the workstation's overall power consumption. The power values are logged and processed in a separate monitor machine (client-server architecture).

Software Setup. Our workstation machine is installed with the latest version of Oracle 11gR2 DBMS under Ubuntu 14.04 LTS with kernel 3.13 to minimize spurious influences, with 8192 as block size. We also disable unnecessary background tasks, clear the system and oracle cache for each query execution. We also disable unnecessary background tasks, clear the system and Oracle DBMS cache for each query execution.

Datasets. We use SSB datasets with a scale factor of 10. It illustrates decision support systems that examine large volumes of data, and execute different types of queries with a high degree of complexity. We have identified the main hierarchies for each dimension table of the SSB multi-dimensional model, applied our formula ($H(Customer) * H(Part) * H(Supplier) * H(Date) = 2^{3-1} * 2^{2-1} * 2^{3-1} * 2^{4-1}$), and generated the resulting 256 possible schemes thanks to attributes correlations. As for workload, we create 30 queries based on SSB datasets, in such a way that two main categories must always be handled: (i) queries with operations that exhaust the system processor (CPU intensive queries) and (ii) queries with exhaustive storage subsystem resource operations (I/O intensive queries). Note that the considered queries include: queries with single table scan, others with multiple joins with different predicates. They also contain sorting/grouping conditions and simple and advanced aggregation functions. These queries are rewritten according to every schema [9].

Evaluation of Scenario 1. As already mentioned, the scenario 1 involved logical optimizations. In our case, we use the default optimizations offered by Oracle 11gR2 DBMS query optimizer. To conduct our experiments, we have deployed the different 256 schemes obtained from varying the initial SSB schema of our $\mathcal{DW}$. The initial queries of our workload are rewritten for each schema (7680 queries all in all) and executed. Execution time (from oracle) and power consumption (from power meter) are recorded. We first analyze one objective function "*power*", depicted in Fig. 12 that confirms power variation according to logical schema and, even better, shows that star schema is far from being the most eco-model. We have noticed that the *co-normalization of the smallest dimension tables* (supplier (2000*SF) and dates (2556) in this case) in the presence of CPU-intensive operations clearly disadvantages power consumption, but neither the number of joins (IO costs) nor the number of CPU-intensive operations (e.g. aggregations/sorting) influence directly the power consumption. A possible explanation is that most of query time execution is spent in CPU processing because data is read quickly due to the files small size. On the opposite, when most of query time execution is spent in waiting until data is ready because of data swapping between memory/disk, less power consumption is recorded.

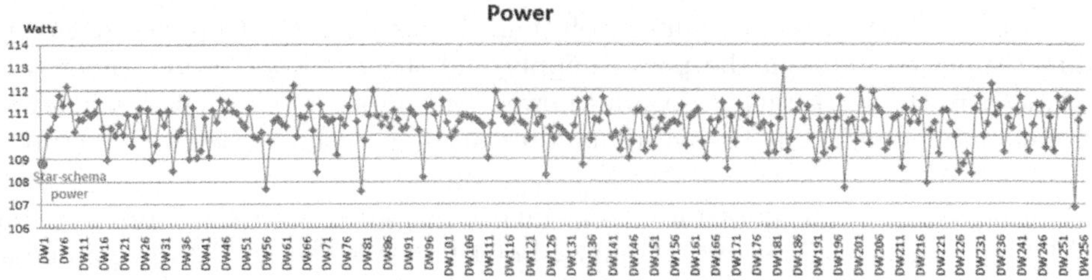

Fig. 12. Impact of logical design on $\mathcal{DW}$ power consumption.

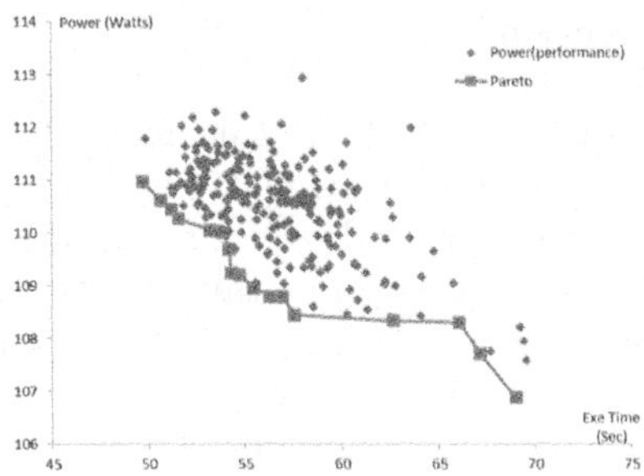

Fig. 13. Impact of logical design on $\mathcal{DW}$ power and performance.

In a second place, we consider two objective functions representing query performance and power consumption. We then highlight the relation between them which takes the form of a convex as illustrated in Fig. 13. This reveals the existence of logical schemes optimizing both NFR (Pareto solutions), and meanwhile, approves our choice of weighting method in selecting schemes.

On the other hand, normalization process reduces storage space, especially with large dimension tables and/or important size of hierarchies. Snowflake schemes are hence appropriate for space-constrained applications as depicted in Fig. 14. This storage gain could be also propagated to storage constraints of optimization structures. These experiments show the limitations of the initial SSB schema to satisfy our fixed NFRs.

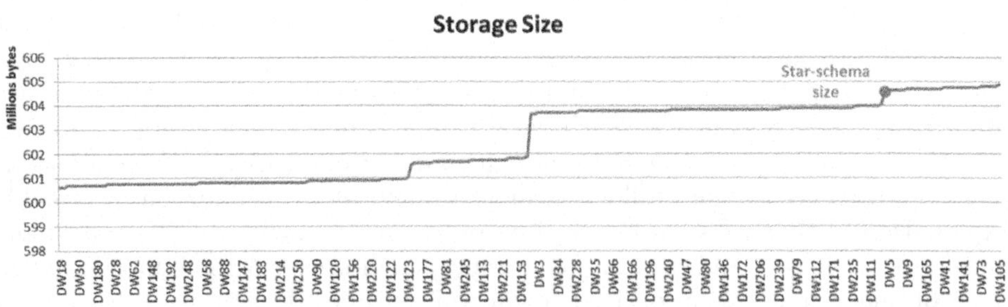

Fig. 14. Impact of logical design on $\mathcal{DW}$ size.

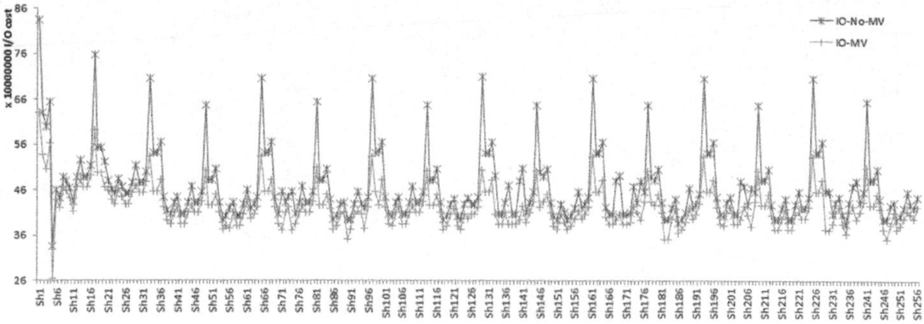

Fig. 15. Impact of $\mathcal{VM}$ on performance of physical optimizations.

Evaluation of Scenario 2. The previous experiments took almost 10 days (7680 queries) what reveals the necessity of using a simulator (cost model) for these and future experiments. We focus in this scenario, on the problem of selecting materialized views by considering the variation of the logical schema (256 schemes), unlike current studies dealing with only one schema. To generalize this, we develop a Java-simulator tool that generates the global plan, using our hypergraph-based approach, for a given workload following any $\mathcal{DW}$ logical schema, and assessed the NFRs cost for the different schemes/workloads using pluggable cost models. Our simulator is equipped with *mathematical cost models estimating different metrics* (query performance, power consumption, etc.) [37]. Figure 15 presents our simulation results of assessing performance (I/O) of the different workloads/schemes with/without views. This attests to the relevance of: (i) materializing views to query performance, which is quite expected. This partially proves the coherence of our cost model, (ii) the impact of logical variability on physical design.

Impact of $\mathcal{VM}$ on both Power and Performance of Physical Optimizations. Rather than testing in combinatorial fashion all the configurations (256 schemes, 30 queries and n views generated by evolutionary algorithms, for each schema), it would make more sense to first select three different logical schemes from our first scenario (that can be hence done using our simulator): a performance, power and trade-off oriented schemes. Using *MOEA*[11] integrated to our tool, we select the set of *Pareto* materializable global plan nodes for each schema. For each materialized view configuration generated by MOEA Framework, the simulator calculates performance and power consumption using cost models. Similarly, it then needs to select a unique view configuration with the desired trade-off: power-MV (*MV for materialized views*), time-MV, trade-off-MV, using weighted sum method with corresponding ω_i. Note that I/O costs were converted to time values (sec).

[11] Java library for multi-objective evolutionary algorithms. www.moeaframework.org.

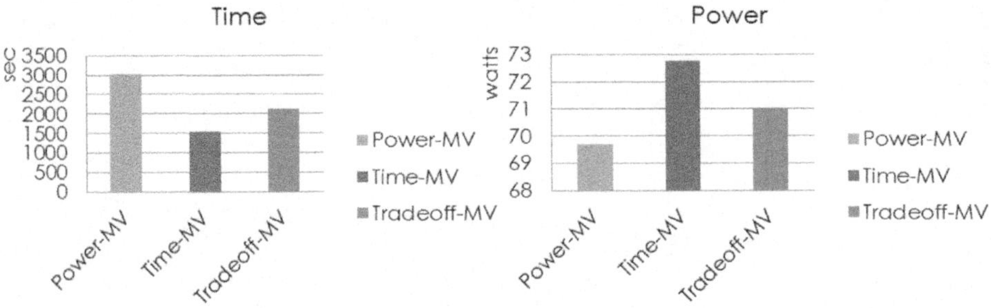

(a) Tradeoff-Oriented logical schema

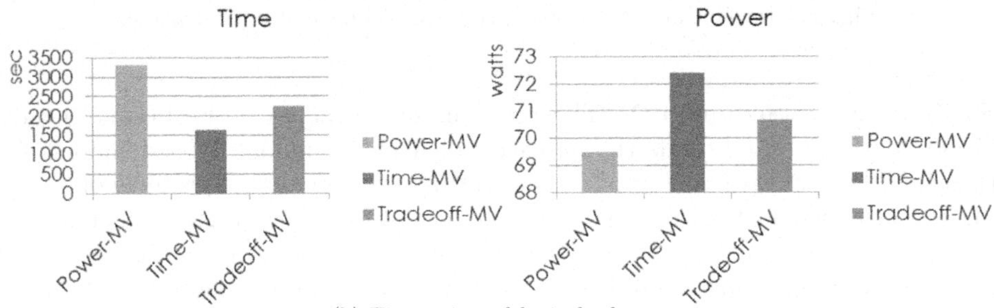

(b) Time-oriented logical schema

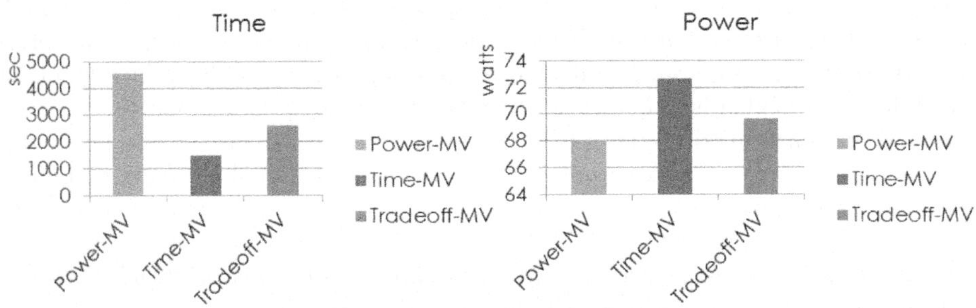

(c) Power-oriented logical schema

Fig. 16. Impact of $\mathcal{VM}$ on both power and performance of physical optimizations.

Our experiments (some of which are depicted in Fig. 16) show that:

- to orient a designer towards a given NFR at earlier stages, she/he must combine the suitable tradeoff of both logical schema and optimization structures. In fact, by choosing a tradeoff-oriented logical schema (Fig. 16-a), we noticed that tradeoff-MV do indeed provide average values of both time and energy. Likewise, by choosing a time-oriented logical scheme (Fig. 16-b), we noticed that time-MV give the best time values compared to other MV (of the same schema), whether they are power or tradeoff oriented. The same applies to the power-oriented schema (Fig. 16-c) where the power-oriented MV do indeed

provide the best power values if compared with time and tradeoff oriented MV (of the same schema).
- logical schemes intended to improve an NFR (time/power), do not necessarily give the optimal values in the presence of optimization structures. That said, they do not give the worst values either. For instance, while it is true that time-VM issued from time-oriented logical schema (Fig. 16-b) provide the best time value if compared with other MV issued from this same schema, this time value is not as optimal as the one obtained by time-MV issued from power-oriented logical schema (Fig. 16-c).
- these results confirm the need for a holistic variability-aware design process where such interdependences have to be considered.

6 Conclusion

In this paper, we discuss a challenging issue related to the integration of energy in the database world. We first summarize the initiatives that the database community did for building energy applications and DBMS. This discussion is based on four actions that we consider relevant which are: *Offer*, *Borrow*, *Reform* and *Evaluate*. According our analysis of the existing studies dealing with integrating the energy in the database technology and the well advanced domains in terms of energy management such as Green Building and Smart Cities, we propose two main initiatives to design existing and new databases/data warehouses. In the case, where a database is operational, we can easily interact with the query processing of the DBMS hosting it. It should be noticed that the query processing takes the lion's share of computation resources. As a consequence, we propose an energy-aware query processing deployed in *PostgreSQL DBMS*. The integration of the energy is performed by the means of the development of mathematical cost models dedicated to estimate the energy consumption when executing a set of queries. At this stage, we have an operational tool, called *EnerQuery*, available at the forge of our laboratory: https://forge.lias-lab.fr/projects/ecoprod and accepted in ACM-CIKM 2016 [39].

The second initiative is to encourage designers to consider the energy when constructing warehouse applications. Due to the complexity of all phases, we consider in this paper the logical phase because it is well connected to the physical one. As a consequence, we launched a think-tank about the impact of varying the logical model of a given $\mathcal{DW}$ on the physical design, according to two non-functional requirements: efficiency of energy consumption and query performance. This think-tank is alimented by **(a)** a debate on the analogy between Software Product Lines (SPL) and $\mathcal{DB}$ design, **(b)** tools to identify/model the dimensions of our problem variability, **(c)** the efforts that designers have to make to deal with this interesting issue. To show the consequences of varying the logical schema on the physical design, we handled two scenarios: **(i)** a physical schema without physical optimization and **(ii)** a physical schema with the process of selecting materialized views. These two scenarios are evaluated using the Star Schema Benchmark and specific hardware to capture energy.

The obtained results shows the worthiness of launching our think-tank that the frozen logical schema, and star schema in particular, is not always the best one to satisfy the fixed non-functional requirements.

Currently, we are working on pushing back the variability to cover other phases of the life cycle such as ETL and conceptual modeling.

References

1. Abadi, D., Agrawal, R., Ailamaki, A., Balazinska, M., Bernstein, P.A., Carey, M.J., Chaudhuri, S., Dean, J., Doan, A., Franklin, M.J., et al.: The beckman report on database research. Commun. ACM **59**(2), 92–99 (2016)
2. Agrawal, R., Srikant, R.: Fast algorithms for mining association rules in large databases. In: VLDB, pp. 487–499 (1994)
3. Anderlik, S., Neumayr, B., Schrefl, M.: Using domain ontologies as semantic dimensions in data warehouses. In: Atzeni, P., Cheung, D., Ram, S. (eds.) ER 2012. LNCS, vol. 7532, pp. 88–101. Springer, Heidelberg (2012). doi:10.1007/978-3-642-34002-4_7
4. Apel, S., Batory, D., Kästner, C., Saake, G.: Feature-Oriented Software Product Lines: Concepts and Implementation. Springer Publishing Company (2013)
5. Appuswamy, R., Olma, M., Ailamaki, A.: Scaling the memory power wall with dram-aware data management. In: Proceedings of the 11th International Workshop on Data Management on New Hardware, p. 3. ACM (2015)
6. Behzadnia, P., Yuan, W., Zeng, B., Tu, Y.-C., Wang, X.: Dynamic power-aware disk storage management in database servers. In: Hartmann, S., Ma, H. (eds.) DEXA 2016. LNCS, vol. 9828, pp. 315–325. Springer, Cham (2016). doi:10.1007/978-3-319-44406-2_25
7. Beloglazov, A., Buyya, R., Lee, Y.C., Zomaya, A., et al.: A taxonomy and survey of energy-efficient data centers and cloud computing systems. Adv. Comput. **82**(2), 47–111 (2011)
8. Bohannon, P., Fan, W., Geerts, F., Jia, X., Kementsietsidis, A.: Conditional functional dependencies for data cleaning. In: ICDE, pp. 746–755 (2007)
9. Bouarar, S., Bellatreche, L., Jean, S., Baron, M.: Do rule-based approaches still make sense in logical data warehouse design? In: Manolopoulos, Y., Trajcevski, G., Kon-Popovska, M. (eds.) ADBIS 2014. LNCS, vol. 8716, pp. 83–96. Springer, Cham (2014). doi:10.1007/978-3-319-10933-6_7
10. Bouarar, S., Bellatreche, L., Roukh, A.: Eco-data warehouse design through logical variability. In: Steffen, B., Baier, C., Brand, M., Eder, J., Hinchey, M., Margaria, T. (eds.) SOFSEM 2017. LNCS, vol. 10139, pp. 436–449. Springer, Cham (2017). doi:10.1007/978-3-319-51963-0_34
11. Boukorca, A., Bellatreche, L., Senouci, S.B., Faget, Z.: Coupling materialized view selection to multi query optimization: Hyper graph approach. IJDWM **11**(2), 62–84 (2015)
12. Brown, P.G., Hass, P.J.: Bhunt: automatic discovery of fuzzy algebraic constraints in relational data. In: Proceedings of the 29th International Conference on Very Large Data Bases, vol. 29, pp. 668–679. VLDB Endowment (2003)
13. Chaudhuri, S., Narasayya, V., Ramamurthy, R.: Estimating progress of execution for SQL queries. In: ACM SIGMOD, pp. 803–814. ACM (2004)

14. Cheong, S.-K., Lim, C., Cho, B.-C.: Database processing performance and energy efficiency evaluation of DDR-SSD and HDD storage system based on the TPC-C. In: 2012 International Conference on Cloud Computing and Social Networking (ICCCSN), pp. 1–3. IEEE (2012)
15. Dannecker, L., Schulze, R., Böhm, M., Lehner, W., Hackenbroich, G.: Context-aware parameter estimation for forecast models in the energy domain. In: Bayard Cushing, J., French, J., Bowers, S. (eds.) SSDBM 2011. LNCS, vol. 6809, pp. 491–508. Springer, Heidelberg (2011). doi:10.1007/978-3-642-22351-8_33
16. Garcia-Molina, H., Ullman, J.D., Widom, J., Systems, D.: The Complete Book, 2nd edn. Prentice Hall Press (2008)
17. Harizopoulos, S., Shah, M., Meza, J., Ranganathan, P.: Energy efficiency: The new holy grail of data management systems research (2009)
18. Hassan, A., Vandierendonck, H., Nikolopoulos, D.S.: Energy-efficient in-memory data stores on hybrid memory hierarchies. In: Proceedings of the 11th International Workshop on Data Management on New Hardware, p. 1. ACM (2015)
19. Hurson, A., Azad, H.: Energy Efficiency in Data Centers and Clouds. Academic Press (2016)
20. Intel and Oracle. Oracle exadata on intel® xeon® processors: Extreme performance for enterprise computing. White paper (2011)
21. Khouri, S.: Cycle de vie sémantique de conception de systèmes de stockage et de manipulation de données. Ph.D. thesis, ISAE-ENSMA and ESI of Algeria, October 2013
22. Kimura, H., Huo, G., Rasin, A., Madden, S., Zdonik, S.: Coradd: correlation aware database designer for materialized views and indexes. PVLDB **3**(1), 1103–1113 (2010)
23. Korkmaz, M., Karyakin, A., Karsten, M., Salem, K.: Towards dynamic green-sizing for database servers. In: International Workshop on Accelerating Data Management Systems Using Modern Processor and Storage Architectures (ADMS), pp. 25–36 (2015)
24. Kunjir, M., Birwa, P.K., Haritsa, J.R.: Peak power plays in database engines. In: EDBT, pp. 444–455. ACM (2012)
25. Lang, W., Kandhan, R., Patel, J.M.: Rethinking query processing for energy efficiency: slowing down to win the race. IEEE Data Eng. Bull. **34**(1), 12–23 (2011)
26. Lang, W., Patel, J.: Towards eco-friendly database management systems. arXiv preprint arXiv:0909.1767 (2009)
27. Levene, M., Loizou, G.: Why is the snowflake schema a good data warehouse design? Inf. Syst. **28**(3), 225–240 (2003)
28. Mami, I., Bellahsene, Z.: A survey of view selection methods. SIGMOD Rec. **41**(1), 20–29 (2012)
29. McCullough, J.C., Agarwal, Y., et al.: Evaluating the effectiveness of model-based power characterization. In: USENIX Annual Technical Conference (2011)
30. Otoo, E., Rotem, D., Tsao, S.-C.: Energy smart management of scientific data. In: Winslett, M. (ed.) SSDBM 2009. LNCS, vol. 5566, pp. 92–109. Springer, Heidelberg (2009). doi:10.1007/978-3-642-02279-1_8
31. Petit, J.-M., Toumani, F., Boulicaut, J.-F., Kouloumdjian, J.: Towards the reverse engineering of denormalized relational databases. In: ICDE, pp. 218–227 (1996)
32. Piatetsky-Shapiro, G.: The optimal selection of secondary indices is NP-complete. ACM SIGMOD Rec. **13**(2), 72–75 (1983)
33. Poess, M., Nambiar, R.O.: Energy cost, the key challenge of today's data centers: a power consumption analysis of tpc-c results. PVLDB **1**(2), 1229–1240 (2008)

34. Rasdorf, W.J., Ulberg, K.J., Baugh, J.W.: A structure-based model of semantic integrity constraints for relational data bases. Eng. Comput. **2**(1), 31–39 (1987)
35. Rodriguez-Martinez, M., Valdivia, H., et al.: Estimating power/energy consumption in database servers. Procedia Comput. Sci. **6**, 112–117 (2011)
36. Rofouei, M., Stathopoulos, T., Ryffel, S., Kaiser, W., Sarrafzadeh, M.: Energy-aware high performance computing with graphic processing units. In: Workshop on Power Aware Computing and System (2008)
37. Roukh, A., Bellatreche, L.: Eco-processing of OLAP complex queries. In: Madria, S., Hara, T. (eds.) DaWaK 2015. LNCS, vol. 9263, pp. 229–242. Springer, Cham (2015). doi:10.1007/978-3-319-22729-0_18
38. Roukh, A., Bellatreche, L., Boukorca, A., Bouarar, S.: Eco-dmw: Eco-design methodology for data warehouses. In: DOLAP, pp. 1–10. ACM (2015)
39. Roukh, A., Bellatreche, L., Ordonez, C.: Enerquery: energy-aware query processing. To appear in ACM CIKM (2016)
40. Royer, K., Bellatreche, L., et al.: One semantic data warehouse fits both electrical vehicle data and their business processes. In: ITSC, pp. 635–640 (2014)
41. Schall, D., Hudlet, V., Härder, T.: Enhancing energy efficiency of database applications using SSDs. In: Proceedings of the Third C* Conference on Computer Science and Software Engineering, pp. 1–9. ACM (2010)
42. Siksnys, L., Thomsen, C., Pedersen, T.B.: MIRABEL DW: managing complex energy data in a smart grid. In: Cuzzocrea, A., Dayal, U. (eds.) DaWaK 2012. LNCS, vol. 7448, pp. 443–457. Springer, Heidelberg (2012). doi:10.1007/978-3-642-32584-7_36
43. Stöhr, T., Märtens, H., Rahm, E.: Multi-dimensional database allocation for parallel data warehouses. In: VLDB, pp. 273–284 (2000)
44. Tu, Y.-C., Wang, X., Zeng, B., Xu, Z.: A system for energy-efficient data management. ACM SIGMOD Rec. **43**(1), 21–26 (2014)
45. Vaisman, A., Zimányi, E., Systems, D.W.: Design and Implementation. Springer, Heidelberg (2014)
46. Woods, L., István, Z., Alonso, G.: Ibex: an intelligent storage engine with support for advanced SQL offloading. Proc. VLDB Endowment **7**(11), 963–974 (2014)
47. Xu, Z., Tu, Y.-C., Wang, X.: Exploring power-performance tradeoffs in database systems. In: ICDE, pp. 485–496 (2010)
48. Xu, Z., Tu, Y.-C., Wang, X.: Dynamic energy estimation of query plans in database systems. In: ICDCS, pp. 83–92. IEEE (2013)
49. Xu, Z., Wang, X., Tu, Y.-C.: Power-aware throughput control for database management systems. In: ICAC, pp. 315–324 (2013)
50. Zhou, A., Qu, B., Li, H., Zhao, S., Suganthan, P.N., Zhang, Q.: Multiobjective evolutionary algorithms: a survey of the state of the art. Swarm Evol. Comput. **1**(1), 32–49 (2011). Elsevier

Author Index

The manufacturer's authorised representative in the EU is Springer Nature Customer Service Centre GmbH, Europaplatz 3, 69115 Heidelberg, Germany. If you have any concerns regarding our products, please contact ProductSafety@springernature.com

Printed and bound by CPI Group (UK) Ltd, Croydon, CR0 4YY
30/09/2026
02225196-0006